PEOPLE HEATERS

A PEOPLE'S GUIDE TO KEEPING WARM IN WINTER

ALEXIS PARKS

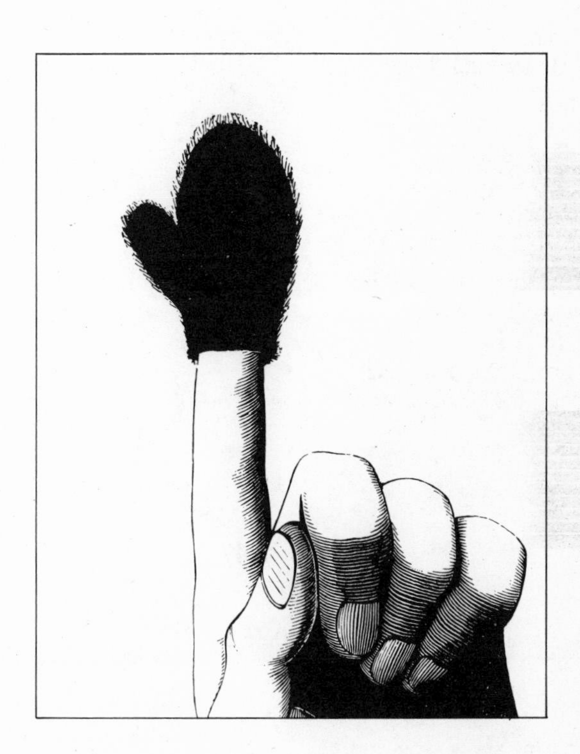

Brick House Publishing Co., Inc.
Andover, Massachusetts

CREDITS:

Excerpt from *The Foxfire Book* by Eliot Wigginton. Copyright © 1968, 1969, 1970, 1971, 1972 by The Foxfire Fund, Inc. Reprinted by permission of Doubleday & Company, Inc.

Excerpt from *Music in the Evening,* copyright © 1960 by Norman Bel Geddes, reprinted by permission of McIntosh and Otis, Inc.

Published by Brick House Publishing Co., Inc.
34 Essex Street
Andover, Massachusetts 01810

PRODUCTION CREDITS

Editor: Jack Howell
Copy editors: Diane Pitochelli and Carol Higgins
Book design: Genesis, Inc./DCP
Cover design: Will Winslow
Production supervision and typesetting:
Dixie Clark Production

Library of Congress Cataloging in Publication Data

Parks, Alexis.
People heaters.

1. Dwellings--Heating and ventilation.
2. Heating. 3. Clothing, Cold weather. I. Title.
TH7226.P29 697 80-21930
ISBN 0-931790-16-6 (pbk.)

Printed in the United States of America

CONTENTS

PREFACE

This was intended as a Winter book, the kind of book that feels comfortable in the company of a pair of long underwear, the warming flame of a woodburning stove, and perhaps a steaming mug of cider.

Those who are drawn to its pages, seeking ways to achieve dramatic savings on their fuel bills will find comfort for the information is there. Those who come to it looking for historic details on how people kept warm in Winter years ago will linger, satisfied, too. Solar enthusiasts will discover that many of the ideas found in these pages should probably be done first before solar designs are applied. And those who find pleasure in innovative and "first of a kind" designs will take heart, for the text includes not only historic but current and future technologies as well.

Finally, for those readers who worry about the steadily rising price for heating fuels and about our country's heavy dependence upon foreign supplies of those fuels, this book will come as a welcome addition to their bookshelf. It can serve as an emergency manual should they ever find themselves facing a shortage or cutoff of those fuels.

Alexis Parks
Boulder, Colorado

This book is dedicated
to my children
Christian and Hillary

Along with the reminder that as a last resort . . . , they can put a match to this book. It's a heat source too!

ACKNOWLEDGEMENTS

This book had its origins in experimentation. The children and I had spent a winter living with the central heating furnace turned "off" at 55 degrees, practicing some of the methods for heating listed in the book. I later described the methods to a friend. Intrigued, he said, "This is something we all need to know how to do. But winter is just around the corner and people are already beginning to worry about how much money they're going to have to pay to heat their homes." Then, knowing that a break was coming up between my writing assignments, he threw out the question, which when answered, hooked me: "How fast can you turn your information into a book?" I began the book that afternoon.

Thus, the first thanks must go to the book's primary and continuing catalyst, Ken Woodcock. Others were instrumental too, for everyone I spoke with while in the process of research, shared not only an immediate interest in the topic but their own methods for keeping warm as well, adding to the information I already had in hand. Thanks also to Dr. Peter Bentley Clark, of Clark-McGlennon Associates, who offered both guidance and a workspace (an office, a phone, and a desk, in his Boston offices as I began my northward journey into the New England states to begin the interview process. To Bruce Anderson, chairman of the Solar Lobby who expanded an initial offer of a 15-minute interview at his home in Harrisville, New Hampshire, into conversations that lasted several days. To Denis Hayes, who took time out to edit the initial draft two days before leaving Washington, D.C. for the directorship of the Solar Energy Research Institute (SERI). To Web Ottis, Director of the Department of Energy's Office of Small Scale Technology and Dr. Jeffrey Milstein, Director of the Office of International Energy Conservation for providing several weeks of essential office space while I researched files; and for their valued expertise and support as well. To others: Christian Herter Jr., Dennis Meadows, Tom English, Joel Schatz,

Amory Lovins, Bill and Sara Bishop (now, like my sister living comfortably in a conventional home with only limited use of their central heating system), Peter Wild, Richard Lopez, Tam Easton, Edward Hall, Dr. Brian Kelly, Dr. A. P. Gagge, Elizabeth and David Dodson-Gray, Ray Manes, Kevin Markey, Mike Joroff, David Gancher, Edward Allen, Buie Seawell, John Muller, David Griffith, Barry Satlow, Mark McCray, Tom Feliu, Dennis Holloway, Bill Hyde, Keith Haggard, Justin Kronwetter, Hans Ueili, Stewart Elliot, Jim Adler, Mike Miller, Doug Baer, James Brookes and others who each, in special ways, contributed to the genesis of the book. To my publisher, Jack Howell, for his steadfast interest and support. And to my family for their love and shared enthusiasm throughout the entire process.

Last winter, a woman in Denver who was stuck on a fixed income and unable to afford the rising cost of home heating, picked up her mattress, dragged it through her apartment to the bathroom and hoisted it into the tub. She went back for blankets, space heater and a small hot plate. She set the hot plate upon a wooden board balanced on the sink and, shutting off the heat in the rest of the apartment, kept warm in the tiny room until spring.

Although the woman avoided freezing to death, an act of such desperation is becoming a sign of the times. We are entering an age of fuel scarcity as well as an age of soaring fuel costs.

"Colder than before" no longer means simply colder than last year, as each winter brings forth its own special candidate for the worst snowstorm in history.

It means a feeling of being cold or actually becoming cold. In adjusting to the higher cost of fuel, we turn the thermostat down. We are living with less heat than before, or living without it.

Today, most of the energy that should be heating a house is wasted energy. Heat is wasted at the place where first produced, wasted in transportation to the homes to be heated and wasted as it escapes through windows, doors, walls and roofs. Lacking social graces, heat takes leave of us at every opportunity. Therefore, trapping it and learning how to focus it requires some transitional steps.

In learning how to conserve expensive fuels and focus heat to keep warm we first need to ask ourselves: do we really need to heat this room or that room when no one is in it? Do we need to heat the entire house just to warm the people in it?

In midwinter, we set the thermostat at a temperature that we assume will be comfortable for us and we end up warming more than ourselves. Our rugs, chairs, walls, pictures hanging on those walls, clothes hanging in our closets and every room in the house will be all heated to the same degree of warmth.

We used to light our houses the same way. Only a few years ago, people thought nothing of leaving on lights in every room of the house, whether there was a person in that room or not. I and most of the people I knew lived that way. Most of us know a few people who still live that way. But over the past decade, as the words "energy" and "conservation" became inextricably linked together, we learned that we do not need to keep lights on in a room when no one is in it.

Likewise, we learned to become more selective in choosing our lighting sources. We choose this light or that lamp for its compatibility to a particular purpose: a reading lamp by a favorite chair, a fluorescent light above the kitchen sink or stove, a low-wattage light bulb for general illumination.

The same selectivity that goes into lighting can be directed toward space heating, by selecting from a variety of methods for keeping warm. We can learn the art of selective heating, choosing items for their thermal qualities, for their heat-retaining and heat-absorbing qualities and thus creating microclimates of comfort around us.

This book is all about learning ways to focus or "localize" heat, to warm people instead of warming the house to heat the people in it, and learning ways to live without central heating systems. It is a smorgasbord of historic and contemporary ideas and methods for keeping warm. The selection, how one goes about keeping warm, can be as individual and as innovative as the person making the choices.

STOVES

AND

"For the right family, a stove still works its subtle ways: it demands that all comfortable chairs face it instead of the TV set; it mysteriously brings the Monopoly sets, the jigsaw puzzles, the battered paperbacks out of storage; it dotes on conversation, popcorn and lingering cups of coffee. . . ." Ann Warner "Tribute to an Old Friend."
Ketchum's Country Journal

Keeping warm with wood has come back into fashion. But burning wood in an open fireplace won't keep you warm. Wood must be burned in an efficient woodburning stove or cookstove, or in a heat-trapping fireplace.

Although our forests could not sustain a wholesale conversion to woodburning stoves—"what are we going to do for shade if we cut all the trees down?"— the change from central heating to localized heating and cooking with a woodstove is happening in those places where a sufficient wood supply makes a woodstove practical.

When selecting a woodstove or cookstove for heating rather than decorative value, you must consider several important factors:

Do you want quick, short-term heat or do you need a slow-burning, sustained fire that will last for hours?

Is your house or central room tight enough so that a small stove will heat it?

Will the fluepipe be used to increase the amount of heat radiated into the room? If so, what safety precautions must be followed?

Is the wood available in your area dry or green and does it have a low or high heat value? [1]

In the mid-1700s the iron stove started out experimentally in this country in the homes of the well-to-do. Because it could provide a more efficient source of heat than the fireplace it soon became the dominant piece of furniture in the room. Unlike the drafty fireplace, the stove could be placed in the center of a room or wherever the heat was needed. The exposed fluepipe radiated additional heat to the room.

The closed, or airtight, stove became vastly improved in the 18th Century, and toward the end of the century it included an oven for heating foods. In the kitchen, the woodburning cookstove was usually placed in front of the fireplace, which was then sealed up, except for an opening to allow the fluepipe access to the chimney. It wasn't long before the nostalgia for the fireplace was replaced by pleasure for the warmth provided by the woodstove. One "Down-East" fellow recalled what it was like when the first cookstove came to his town.

"I recollect as well as if 'twas yesterday when the old man fetched home the first stove we ever had. He brought it up to Portland (Maine), and I tell you it wa'n't much like the stoves we have now. It come knocked down, as you might say, and being as it was the first stove anybody hereabouts had ever seen, there wa'n't anybody knowed how to put it together. The man that sold it to my father told him the bolts mus'n't be screwed up tight but must be set up gradually as the stove het up or the iron was likely to crack. Well, we couldn't seem to get it together nohow, so at last one of the boys went up to the Center and got the blacksmith to come down and help us, and after a while we got it set up."

"The old folks missed the fireplace, and for a while they thought they'd freeze. Nobody knowed how to build a fire in a stove or how to keep it going once it was built, and the old woman couldn't seem to get the hang of cooking by it, but after a spell we got used to it and the old man had all the fireplaces bricked up and bought stoves to heat with; little sheet-iron stoves, they was, but they het up quick and didn't burn nigh the wood a fireplace did."

By closing off the central fireplace, they were able to seal off a source of steady, cooling drafts. The woodstove, as they finally discovered, provided a means of heating and, ultimately, cooking which was both more effective than the fireplace and burned less woodfuel.

In the mid 1800s 90 percent of home heating needs were supplied by wood. By 1950, although one-third of the world still used wood as its primary heating source, only 3 percent of the homes in North America were heated that way.

Today the percentage of homes heated by woodburning stoves is on the increase. This increased use is especially noticeable in the Northeast where wood supplies are plentiful. Vermont, for example, produces 10 percent of its electric power needs by burning wood chips and wastes. According to a report by the New England Office of the Department of Energy, 50 percent of the fuel consumed in Maine could be produced by wood by 1985. However, Bruce Anderson, Chairman of the Solar Lobby in Washington, D.C., reminds us, "after the first cord, the novelty wears off. Most of the houses that are designed to maximize the use of solar energy don't require more than a cord. Then wood remains a pleasure to use" [2].

Many books cataloguing the varieties and benefits of wood-burning stoves are available. Also, many books describe which types of wood are best to use. This section does not attempt to duplicate those efforts. It simply draws important distinctions between stoves and between types of wood in terms of their value as heat sources.

People who own woodstoves know that strange dependency grows between themselves and the stove. It becomes a part of the family, demanding its share of attention along with everyone else. "It has a voracious appetite and the ability to perform magnificently, but does pretty much what it damn

pleases. It can make you glow with warmth and affection, scare the stuffing out of you when it really gets going and make you mad enough to swear and kick" [3].

Still, after the swearing and kicking are done, the durable woodstove continues doing what it does best: providing a ready source of radiant heat.

Like the fireplace of old, the stove also provides a focal point for many activities. Those who heat with a woodburning stove—by choice—may find the dispersed heat of a central heating system disconcerting. One old-timer found himself pacing the floor at his son-in-law's place. "I didn't know where I was supposed to sit. Everything was the same temperature."

Several variables determine which type of wood stove is the best. One variable is construction of a house. If a house is poorly insulated, a large stove may be necessary to heat it. If it is well sealed, a small stove may suffice. In some passive solar designs which emphasize gain from the sun and tight draft-proof construction, the problem is to provide supplementary, but not excessive heat.

Woodburning stoves are usually divided into four design classes:

- Airtight - in which the firebox can be closed off entirely
- Front-end Burner - with baffles that direct the airflow from front to back
- Convertible - which can be left open for a view of the fire or closed down to become airtight
- Downdraft - an airtight stove which draws down the primary air from the top then recirculates it through a series of horizontal baffles

For the purposes of this section these stoves have been regrouped into categories according to heat sink (heat sponge) values rather than design class.

Most woodstoves sold today are lightweight, portable models which offer only quick, short-term heat. Since they have thin walls, they are prone to warp, bend or crack more readily than heavier–metal stoves. They will also rust more quickly in damp areas of the country. But if all you need is something to take the chill out of the air inexpensively and quickly, then the lightweight woodburning stoves will do well.

For long-lasting heat, the following types of stoves should be considered: heavy plate or cast-iron stoves, ceramic tile stoves, soapstone stoves and stoves with heat pipes, such as the SuperJuca. Two low-cost stoves are also considered in this section: the SEVCA, which costs only a few hundred dollars (less for low income folks) and consistently comes out on top in stove ratings; and the Yukon barrel stove.

One woodstove, the classic soapstone stove built in New England in the 1890s and reappearing in the region today, is made of blocks of soft talc rock called soapstone [4]. Soapstone is especially good at absorbing heat from burning wood and releasing it slowly into the room.

Blocks of soapstone once served as foot and bedwarmers. They could be heated on a stove—each family member had one—and when wrapped in a cloth could be carried to bed to warm the sheets. They serve the same long-lasting heating function when attached to the sides of a metal stove.

One disadvantage of soapstone and ceramic tile stoves is that they are slow to warm up. But once warm, they hold and radiate heat 6 to 12 hours after the fire has burned down.

Weight can also be a problem. Because these stoves typically weigh more than most cast-iron and steel stoves, they require a special fireplace-style foundation. But for a quality stove that mixes beauty with airtight, front-end burner efficiency, few can match these classic stoves.

Ceramic tile stoves do not have a "burnt metal" smell like some cast-iron and steel stoves and can be a work of art as well as efficient. They were first invented in 13th-century Germany, and come in a variety of solid color or handpainted tiles.

Despite knowledge of the ceramic tile stove's efficiency, most New England settlers continued to cling to the traditional English method of heating their homes with fireplaces well into the 1800s. While fireplaces draw off most of the heat generated by by the fire—between 80 and 90 percent—and burn primarily wood products, ceramic stoves give back most of the heat to the room, and can burn coal, wood or peat. Some can even be converted to use oil, gas or other solid fuels.

Another type of woodburning stove which, like the ceramic tile and soapstone models, offers long-lasting radiated heat is the cast-iron or heavy-steel plate stove. Like the others, these heavy-metal stoves are typically lined with firebrick to increase the stove's efficiency. The firebrick helps keep the temperature high inside the stove's firebox and serves as a heat sponge to absorb and slowly release the heat. Because these stoves are made of heavier metals than the typical woodburning stove, they are more durable and less prone to warping and cracking than the light-weight models. However, these problems are not entirely eliminated, and most wood heating guide books recommend that you follow the owner's manual closely when firing up a new stove.

Some of the stoves that provide a heavy thermal mass in the room for long-lasting, radiated heat also provide a view of the fire through a fireproof glass window. Another style is the popular, though less efficient, "Franklin" (convertible) woodstove. The visible flame, long a source of poetry and romance can also, when low, serve as a reminder that more logs need to be added. But if you want to maximize the radiated heat while at the same time controlling combustion, stoves that remain closed and are surrounded on all sides by heavy metal, tile, or soapstone may be the best.

In all airtight stoves, regulating oxygen intake through an opening in the front or top of the stove controls the rate of burning. The longer the burn from a single load of fuel, the longer the heat from that fuel is radiated into the room.

The Tempwood stove, for example, draws air into the stove from the top. The fire is set in reverse—with the kindling on top instead of the bottom. The effect of the air drawn from the top to the bottom is like a blowtorch. In addition, as the air rises back through the flame, there is, in effect, double combustion of gases.

The airtight Jotul stove from Oslo, Norway—like all front-end burners or downdrafters—includes a series of baffles. The heat must pass back and forth through the baffles before leaving the stove. In the process of slowing down the departure of the gases, the Jotul introduces a secondary source of air to insure double combustion and thus reduce the creosote buildup in the fluepipe. This type of stove is designed to extract as much heat as possible from the burning logs.

The JUCA Super-Fireplace is different from the others. It is a fireproof, glass-enclosed, free-standing fireplace. It includes a unique series of 3-inch-long pipes installed above the stove body and hidden from view beneath a tent-shaped hood. The pipes are connected to a sleeve that runs up the back and around the sides of the fireplace. When a small blower is turned on, the heat is drawn through the sleeve, through the pipes and down onto the floor. Heat ducts can also be attached to direct the hot air to other parts of the house.

This half-serious balance sheet on the costs of converting from oil heat to a wood stove was compiled by a Vermonter who apparently has a sore back and a broken home.

Economic Justification of Woodburning Over Fuel Oil—First Year Costs

Item	Credit	Cost
(2) Stoves, equipped and set up	$—	$1,385
Remove hot water baseboard system w/boiler	—	238
Sell hot water baseboard system w/boiler	125	—
Pursue reputable wood dealer (not available)	—	76
Buy: Chain saw	—	210
Axe, hatchet, wedges, maul, cant hook	—	119
Old truck (scrapped after first load)	—	595
New 4-wheel-drive truck	—	8,645
Wheel chains	—	88
Replace truck rear window (twice)	—	310
Work gloves (swiped from shop)	—	—
Fine for cutting wrong trees	—	500
Buy: 5-acre wood lot	—	4,995
Splitting machine	—	950
14 cases Miller beer	—	126
6 fifths ginger brandy	—	38
Fine for littering	—	250
Towing charge (creekbed to hardtop road)	—	50
Gas, oil, chain sharpening and Band-Aids	—	97
Doctor's fee for splinter removal from eye	—	45
Safety glasses (swiped from shop)	—	—
Medical costs for broken toe (dropped log)	—	128
Safety shoes	—	35
Repair burned hole in living room carpet (unsuccessful)	—	76
New living room carpet	—	699
Paint living room walls and ceiling	—	110
Taxes on wood lot	—	44
Wood lot boundary dispute settlement	—	469
Roof repair after chimney fire	—	840
Fine for assaulting fireman	—	50
Extension ladder	—	55
Chimney brush	—	22
Medical cost for broken leg (fell off roof)	—	478
Chimney cleaner service	—	90
Coffee table replacement (chopped up and burned while too drunk to bring up firewood from cellar)	—	79
Divorce settlement	—	14,500
Annual Fuel Oil Saving	376	—
	$501	$36,388

Net Cost of First Year Woodburning Operation, $35,887

Source: "Economic Justification of Woodburning Over Fuel Oil," or "What It Costs to Own a Woodburning Stove," Outlook Section, *The Washington Post,* April 1, 1979.

While several types of stoves have been mentioned, a number of other, equally good stoves of similar design and quality are now on the market. However, it is important to realize that not all woodstoves are necessarily good stoves.

Some so-called closed stoves are not airtight and burn wood so quickly that they must be continually refueled to maintain their radiant heating capabilities. A lightweight, thin-walled stove may burn through or crack with continued high heat. Replacing it may ultimately be more expensive than buying a heavy-weight stove that could outlive us by centuries. The hidden costs of inexpensive stoves may not be known until after installation.

Part of the problem with low-cost, poorly designed wood stoves is historic. When woodstoves first became popular in the early 1800s, it took a lot of refinement and a careful study of stove efficiency to come up with a stove that did what was expected of it. Stove manufacturers today are having to relearn a nearly forgotten art.

The Franklin stove is a good example of this. The Franklin that we are familiar with today bears little resemblance to the cast-iron stove that Ben Franklin invented in 1739-40. Franklin did not patent his stove, but gave the design to a number of people with the hope that the stove would be widely distributed. To trim costs and to fit the stove inside existing fireplaces, a number of changes such as elimination of the baffled air chamber were made. These changes ultimately stripped the Franklin stove of all its useful benefits. However, these inferior stoves continue to bear a resemblance to Franklin's design, and thus continue to carry the name of the famous Pennsylvania inventor.

There is one stove that combines quality and low cost. What it lacks in good looks, it makes up for in efficiency.

The SEVCA stove [6], an acronym for Southeastern Vermont Community Action, has received top ratings in stove efficiency tests. The stove, designed and produced by the Vermont anti-poverty agency, is made from a discarded propane gas tank. Inside the tank are two chambers: a combustion chamber and, above that, a smoke chamber. The upper chamber—like an extended fluepipe—permits the extra heat from the fire to be retained rather than lost to the out-of-doors.

The Yukon stove, like the SEVCA, is a low-cost stove. It is a double-decker wood-burning stove which can include a smoke exchanger with a built-in oven.

This stove, made from discarded 55-gallon oil drums, can be set horizontally or vertically with a door cut into the side or the end. It combines do-it-yourself ingenuity, recycled materials and low cost. Although the Yukon stove has all the draw-backs of lightweight stoves, it can be adapted to include some of the advantages of the heavier models. The interior can be lined with firebrick, furnace clay or stoveliner to protect the sides from high temperatures and to provide extra storage for the heat. A firegrate can be used to lift the burning logs off the bottom. The overhead oil drum will capture and radiate extra heat as the smoke rises up the fluepipe [7].

In the 1970s there was almost no mention of cookstoves in the popular literature. One found cookstoves listed in antique catalogs and sold in antique stores. Now, new woodburning cookstoves are making their way back to this country from Europe.

There are still a few people around who remember what growing up with a cookstove was like. "All the years we were growing up," one of them recalls, "there was not one winter morning when we did not race downstairs from our icy bedrooms to find the kitchen stove giving off its warmest glow, the coffee made, the teakettle already simmering, for one of our parents was always up before us to perform this service of love. There was a huge double-boiler of oatmeal that had been made the night before and set toward the back of the stove to cook slowly to just the right flavor and consistency—a creamy masterpiece that was served with brown sugar and top milk. . ."

"For children, the kitchen stove was like a third parent; we dressed and undressed beside it, took our baths in the old washtub pulled into its halo of warmth, did our homework on frigid nights with the oven door open and our feet propped inside. We learned to cook early in life, because it was fun and we loved good food, and because on a winter day or rainy night it was the coziest thing to do" [8].

Cookstoves can be multipurpose. They can cook the food, heat hot water for domestic use, and radiate heat to the kitchen. But if you have little experience with wood cookstoves you should probably not buy one for multiple use. Stoves used for both cooking and central heating require a lot of attention.

Learning the knack of cooking with wood heat isn't easy. Maintenance of a "steady" oven temperature requires constant attention to fueling the stove.

The cooking surface does not heat uniformly. If the top has several "eyes" each has a separate temperature. The eye closest to the fluepipe at the center back is the hottest.

Because the heat cannot be regulated, the pots must continually be moved around according to how much heat is needed. If heat is needed simply to warm a room, it might be more easily provided by a simple, efficient woodstove. But those willing to try a cookstove now have a welcome choice. The best new cookstoves on the market include the Lange, made in Denmark, the Stanley, made in Ireland, and the Austrian-made Styria, the Rolls Royce of cookstoves.

The Stanley Mark 1 has the classic old-fashioned " as American as apple pie" look. This is fine on a cold winter day but may create too much heat in the summertime. Black paint or white enamel models run $895-$995 (1979 prices).

The Lange 911-W comes in a variety of enamel colors—dark green, cobalt blue, cranberry or black— or in black paint. It is an excellent space heater as well as a cookstove and sells for $725-$875.

The Styria, the most expensive of the cookstoves offers superior quality and superior materials... for a price ($1000-$2000).

The English-made Rayburn cookstove ($1800) is not exactly a space heater. Made by the same company that manufactures the famous AGA cooker, it is so well insulated that heat does not radiate into the room. Instead, hot water heated in the stove can be piped from the stove to steam heat two rooms. Interestling, the AGA cooker, invented in 1924, was the "perfect solid cooker of its time." It depended on excellent thermal insulation, Swedish iron and "sound principles of physics." It was developed by Nobel prize -winning Swedish physicist Gustav Dalen after he lost his eyesight while experimenting with cookers.

Before ending the discussion of woodburning cookstoves and stoves as a source of heat, I must mention fluepipes.

There are few government regulations for installing woodstoves and fluepipes. Laws vary from state to state and sometimes from town to town. There are even varying rules of thumb regarding how close a stove can be to a wall and how long the fluepipe should be.

Some users feel the fluepipe should be as short as possible for safety reasons. Some insurance carriers require that the fluepipe be short. Other users create elaborate angles and extensions, or attach hot water units or blowers to the pipe to increase the amount of heat radiated from the stove into the room. A 55-gallon barrel of water, for example, can be inserted on the pipeline (perhaps located on the second floor if the stove is placed on the main level.) The water will serve as a storage medium for the heat but the barrel must be clean and leakproof.

Some woodstove users prefer to bend the fluepipe into several angles or into a "donut" shape above the stove.

A typical length of four to six feet of fluepipe contributes 15 to 25 percent of the total heat from the woodstove. Adding five more feet increases the efficiency by nearly 8 percent.

Creosote buildup may occur if the pipe is too long and the flue gases have a chance to cool down. Creosote, the unburned wood-tar vapors that attach to water vapor released by greenwood, will cool and condense on the inside wall of the fluepipe. The problem can be reduced by insulating the pipe or chimney and by using dry wood; however, in a modern, well-insulated house, an extended fluepipe may not really be necessary.

The Shaker community, one of the single most inventive groups of people living in America 100 years ago, preferred the extended fluepipe. With the woodstove placed on one side of a room,

the fluepipe stretched across the entire length of the room, hanging about two feet down from the ceiling, or wrapping its way up around the walls of a room, exiting near the ceiling.

After the Shakers discovered how efficient the woodburning stove was, they sealed up every fireplace in their buildings and left only a small hole for the fluepipe to enter the chimney. Their small stoves were close to the floor and the pipe was extended to a practical length.

In recent years, the National Fire Protection Association (NFPA) has sought to standardize woodstove installations to reduce the risk of fire. The NFPA believes that pipes connecting the stove to the outside vent should be as short as possible and recommends a maximum of 36 inches. The distance between the stove and the wall can be reduced to 18 inches by using insulating materials such as asbestos, or sheet metal held 1 inch away from the wall by noncombustible spacers, or by using a wall of fireproof bricks.

However, a fireproof backing is necessary whether the fluepipe is long or short. A woodstove generates a substantial amount of heat and may set fire to nearby combustible materials such as curtains, a wood floor or even wood beams inside an unprotected wall. (A fire extinguisher nearby, though not required, can help reduce the risk of accidental fire damage.)

To provide a heat "sink" fireproof bricks can be placed beneath and behind the stove to absorb and slowly release the heat. A piece of sheet metal will reflect heat. Some people want only an extra boost of reflected heat, while others want to hold the heat in the room longer, and thus use bricks.

The science of stove efficiency was a popular topic in polite conversations during the 19th century. Though most people had learned the peculiarities of the woodstove by experience, some of them could turn to manuals, such as the one by Catherine Beecher and Harriet Beecher Stowe, for practical guidance. The problems with chimneys and fluepipes, as well as solutions, were listed simply. Today we are also learning by experience, but there are guidebooks to help direct us too.

As mentioned at the beginning of the chapter, a fireplace won't keep you warm unless it is enclosed. A fireplace insert will provide some additional heat but to be most effective, the insert must sit in front of the fireplace, where it releases heat into the room instead of up the chimney. Of course, a fan can be attached to help distribute the heat whether the stove is set in front of or inside the fireplace. And to avoid drawing air that has just been heated back in to the stove, an outlet can be installed to draw in cold air from the outside for combustion directly inside the insert or the fireplace.

The fireplace can also be made to serve as a heat sink. The Russian fireplace does this with a series of baffles in the masonry through which the smoke and heat must weave before leaving the chimney. As in the wood-burning stove with a built-in baffle system, the longer the smoke and heat take to get up the chimney, the more heat is absorbed and released by the masonry. The excess heat can also be directed into a rock storage unit for later release into the home heating system.

The Inglenook, used for centuries in parts of northern Europe, is similar to the Russian fireplace. It, too, uses masonry to absorb and radiate heat from a fire, but it is built in the shape of a dome or "beehive." Members of the household sit inside this small room—literally a room-within-a-room—and are warmed by the heat radiating from the fireplace and the surrounding walls. The walls, in time, radiate their warmth to the rooms surrounding the domed structure.

Most long-time wood users will tell you that the best wood for burning is dry wood—wood cut and stacked a year before it is burned. But most people won't mention that some kinds of wood produce nearly twice as much heat as others.

Hardwoods, such as hickory, oak, elm, birch and cherry, among others, produce nearly twice as many Btus (British Thermal Unit) compared to softwoods. (A Btu is the amount of heat it takes to raise a pound of water one degree Fahrenheit.) Hickory produces 24,500,000 Btus per cord. Softwoods such as cedar, poplar, spruce and pine produce only half as much heat but are easy to split. Because they are easy to cut they are good for kindling.

Most wood users prefer to stack their logs in a well-ordered pile measuring four feet by four feet by eight feet, the standard size for a single cord of wood. The wood is stacked off the ground on a platform comprised of two wood "runners." The rows are stacked in alternating directions, allowing as much air as possible to circulate around the logs and hastening drying.

Another method from Canada, which may speed the drying process, is to select a spot in the yard where a "stump" can be set and stack the cut wood vertically, in "tepee" fashion, around it.

REFERENCES AND NOTES

1. Nathan C. Fuller, The Down East Reader, Selections from the Magazine of Maine (Philadelphia: Lippincott, 1962), p. 28.
2. Private conversation with Bruce Anderson, recorded prior to his acceptance of the chairmanship of the Solar Lobby in Washington, D.C. He was speaking, at the time, of homes designed by Total Environmental Action of which he is President.
3. and 8. "Tribute to an Old Friend" by Ann Warner. Reprinted by permission from Blair & Ketchum's Country Journal. Copyright © 1976 Country Journal Publishing Co., Inc.
4. Woodstock Soapstone Co., Inc., of Bridgewater, VT 05034 is one of several companies manufacturing soapstone stoves.
5. For more information contact: JUCA, Inc., Dept. ME, P.O. Box 68, North Judson, IN 46366.
6. SEVCA, Southeastern Vermont Community Action, Box 396, Bellows Falls, VT 05101
7. David Haven, *Woodburners' Handbook*, (Portland, ME, Media House Publishers, 1973).

SPOT

HEATING

"To keep warm, I'd shut down two-thirds of the house, put multiples in bed, put a wool hat on my head, a scarf around my neck, wear my long underwear and . . . I guess, keep my hands in my pockets." Dennis Meadows

Central heating systems are typically designed to meet the family's "average" heating requirements. But if the thermostat is lowered a few degrees or if the heat is localized and used only where needed, then the central system can be considered oversized, too large to meet our limited needs and should be scaled down. If the house is designed or redesigned to emphasize the best passive solar techniques you might not need a heat distribution system at all because the house would lose the heat so slowly.

If we learn to focus the heat and use it only where and when we need it—warming just the people in the room and not the room itself—our so-called average heating requirements could become very small indeed. We would, in other words, have to redefine that average.

There are many sources of focused or "spot" heating in addition to the familiar woodburning stove. Most spot heaters, such as the electric coil that heats a single cup of water or the electric space heater, need modest amounts of electricity to make them work. Others, such as the old-fashioned footwarmer, the "heat-grabber" (a window-box device that provides warm air) and futuristic eutectic tiles, do not.

Each, in their own way, provides a spot of warmth. Some can be carried from place to place while others cannot.

The portable electric space heater, for example, can be carried from room to room, from work to leisure area. A heating unit can be attached to the walls, incorporated into the wall or hung from the ceiling. If large enough to heat several rooms, a space heater can be part of the permanent room furniture.

A localized heat source creates a pocket of comfort. A hundred years ago, this pocket of comfort could be taken literally. Children were sent off to school with hot potatoes or hard-boiled eggs in their pockets. These portable heat sources were also part of their lunches.

For city dwellers, a spot of comfort was often the corner hot chestnut stand, the baked-potato man's cart or the coffee stall. "Before six in the morning," noted the illustrator Gustave Dore a hundred years ago (after his pilgrimage through the streets of working-class London), "while the mantle of night still lies over the sloppy streets, and the air stings the limbs to the marrow; the shadows of men and boys may be seen, black objects against the deep gloom, gliding out of the side-streets to the main thoroughfares. . . .The baked-potato man and the keeper of the coffee stall are their most welcome friends—and their truest; for they sell warmth that sustains and does not poison" [1].

Sustaining warmth could also be carried in small pieces of soapstone. When heated on a wood-burning stove, wrapped in newspaper and then in cloth, these soft talc stones would radiate warmth for the hands or feet for hours. Larger blocks of soapstone and sad irons (sad meaning "heavy"), which were less portable because of their weight, were used to warm the hands and feet at work or home. Footwarmers, the small metallic boxes designed to carry lumps of hot coal away from the fire, were a fancier and more portable way of providing the same function.

Another form of portable heating was introduced after the discovery of oil in Pennsylvania in 1859. A number of small oil, kerosene, paraffin (also a petroleum product) and gas heaters came into popular use. The first cast-iron radiant heater using a gas flame and tufted asbestos placed between fireclay bricks was made in 1882.

Electric heaters, whose portable "fires" could be turned on and off whenever desired came into use shortly after that. However, the real breakthrough in electric space heating did not occur until 1906. In that year, the wire element of one space heater model was made of rustproof nickel and chrome and, for the first time, was unaffected by continuous heating and cooling or by moisture.

In time, these small space heaters, like the woodburning and coal-burning stoves and furnaces, gave way to forced air and steam central heating systems. Although some of the older housing units in this country still depend on space heaters for heating, they have, for the most part, become space holders, stored in the attic or closet, disposed of or given away. However, in many parts of Europe where heating fuel is in limited supply and highly priced, space heaters are commonplace.

A number of low-cost, portable space heaters are reappearing on the market today. As before, their utility lies in the fact that they can quickly and inexpensively take the chill out of a room that is used only occasionally or has limited use. If the general heat in the house is turned down or off, if you rise early or go to bed late and are looking for immediate or focused warmth in several areas around the house, the electric space heater can provide that warmth wherever and whenever needed.

Most electric or propane space heaters are sold as either upright or baseboard units. Because propane space heaters will consume the oxygen in a room, they are usually used in an open, airy space such as an unheated garage. In the airport terminal in Gillette, Wyoming, an infrared heater suspended from the ceiling over the rows of seats in the passenger waiting area radiates heat downward. Because it is difficult to maintain a uniform level of heat inside the building, the general traffic areas are left cooler and supplemental heat is provided where people congregate.

Electric space heaters and new, oil-filled, portable electric radiators [2] are most commonly used in home heating. The heat is either circulated into the room by convection and assisted by a fan which disperses the warm air, or radiated into the room from a reflective shield placed behind the heating elements. The radiated heat mixes with the cooler air and, warming it, rises in the room.

Because the convection-only models, like central heating systems, do not warm directly, but indirectly by warming the air, a combination convection-radiant heater is usually best for heating a room.

The choice between upright or baseboard models can be made according to convenience or need for portability. Baseboard units can be moved from room to room but are more awkward to carry than the upright heaters. An architectural designer who built his own solar home made interesting use of baseboard units. He had planned to heat his house by sunlight and a supplemental woodburning stove, but the bank which loaned him the money for construction required a conventional "back-up" system.

To meet the bank's requirement, he installed plug-in baseboard heaters. These satisfied the bank's request, but because they were left unplugged, supported the owner's desire that the house be 100 percent solar heated.

Because they are not problem-free, space heaters deserve a few notes of caution. If not properly handled, a space heater can set nearby combustible materials on fire, cause serious burns or release noxious gas fumes. Because of this, space heaters should be kept away from small children and away from draperies and other flammable materials. Older children should be cautioned not to touch the heater or to let flammable clothing hang close enough to catch on fire. The heater should also have a thermostat control and an automatic shut-off valve in case it is accidentally knocked over.

In fuel-conscious Germany, a large, nighttime, rock-storage space heater or "Nachtsteicher-heizing" takes advantage of the inexpensive middle-of-the-night (off-peak) electric rates offered by the utilities. The German heater uses electricity to heat the rocks during the night hours.

The unit is automatically turned on between 10 p.m. and 6 a.m. and the stones are heated to 150 degrees centigrade or higher. A fan disperses the stored heat into the room during the day. If necessary, the unit can be given another boost during off-peak daytime hours.

Although the night storage system has been in use in Germany for some time, the idea has been gathering limited interest in this country since the late 1970s. Those who promote experimental use of such a system have sought to incorporate it into the heat distribution system of the central furnace. The unit, called a heat storage furnace, would store enough heat in an 8-hour period to provide 16 hours of hot water and space heating.

"The heart of the heat-storage furnace is a core formed of a stack of high-magnesium-content brick. During periods of charging, the ceramic material is heated by ceramic-sheathed, open-coil, nichrome, low-watt-density elements, which slip into the recesses molded into the bricks" [3]. The heat in the storage unit will reach temperatures around 1400 degrees Fahrenheit when it is fully charged.

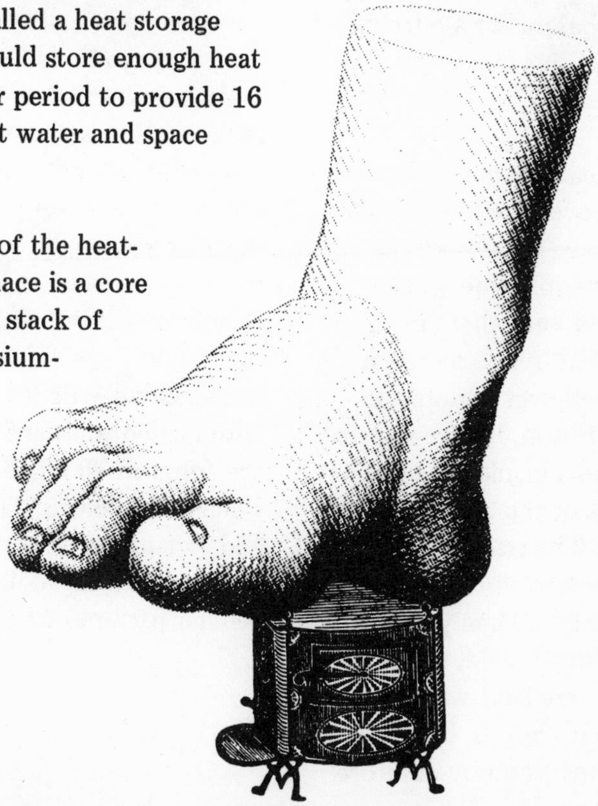

One major difference between the storage unit in operation in Germany and the thermal storage furnace just described is that the latter ties into the central heating air duct system while the German space heater becomes part of the permanent furniture in the room it heats. Another important difference, of course, is the inexpensive off-peak rate charged by the German utilities to owners of nighttime, rock-storage units.

The Japanese use another type of spot heating. In some of their airy, thin-walled houses, a space heater is placed in a hole cut into the center of one room. The hole measures about two feet by three feet square. If the room is a four-tatami-mat size (each mat measuring three by six feet) the square is cut where the four mats come together. The space heater can be an electric radiant heater, a hot plate, a hibachi that burns smokeless charcoal or a burning ember placed in a closed pan and wrapped in cloth.

The small heater is placed in the hole, then covered with a wood screen or guard to protect the feet of the people who will sit around it and keep them from getting burned by the metal. A table is placed over the hole, a blanket, or Futon, is draped over the table and a blanket is pulled over the legs of the people sitting at the table. The traditional Japanese custom is to sit close to the floor. Therefore, when people sit around the small table, their legs and feet are kept warm by the small space heater. Their backs are kept warm by comfortable quilted robes.

Because many Japanese houses are quite open and airy, smokeless charcoal-fueled hibachis can be safely used. However, in the typical American home, using this type of heater in a closed room might prove fatal to the occupants. Smokeless charcoal, even when covered with sand, emits odorless, deadly carbon monoxide as long as the coals are burning.

One unique kind of radiant heater that differs from the standard electric space heater is the heated panel or wired wallboard. Electric wires set into a material like gypsum board can generate a low-grade, localized heat. Unlike microwave, the panel produces a longwave radiation similar to heat from a radiator. If it is placed on a wall, it radiates warmth like a hot water bottle. When it is placed in a high ceiling, however, it may take an hour or longer to radiate heat down to the floor.

Frank Lloyd Wright experimented with heated walls and ceilings at the Johnson Wax Laboratory in Racine, Wisconsin. All the intake and exhaust pipes and the utility lines were arranged like the cellular pattern of a tree trunk.

Each ran in its own groove in the concrete floors which then joined together in the vertical stack. In this manner, low-grade, localized heat was distributed throughout the entire building.

Earlier in the century, the architect Norman Bel Geddes designed an electromagnetic chair which provided low-grade heat to the person sitting in it. Today we can purchase works of art that provide the same function. The Aztec Marketing Company [4] sells a 24-inch by 36-inch sand-coated decorator panel. The panel can be purchased plain or with one of seven painted designs. The panel uses only about 500 watts of electricity or about one third the wattage of a conventional space heater, and heats up to about 180 to 200 degrees Fahrenheit. A smaller, 200-watt picture-size panel, according to the literature, will keep a person warm at their desk. While both sizes of radiant heaters never get hot enough to burn the skin if touched, bumpers on the back of each picture lift it off the wall. This allows air to circulate behind the panel and keep wallpaper or paint from being affected by the heat.

In addition to the various methods of space heating previously mentioned, there are a number of new, nearly new or "rediscovered" inventions appearing on the market. These products offer a unique selection of spot heating devices to provide extra heat in the winter, such as the electric floor heating mat [5]. The mat, which uses less electricity than a 100-watt bulb can be used to keep the feet warm while sitting at a desk or standing behind a counter.

Perhaps most unexpected is the convection space heater that is being designed to use the waste heat from an ordinary refrigerator motor. Refrigerators are notorious energy wasters. In fact much of the need for cooling in a refrigerator is created by excess heat from the refrigerator motor as well as the loss of cool air through poorly insulated doors. "If all refrigerators were made only 50 percent more efficient," Denis Hayes, Director of the Solar Energy Institute once pointed out, "they would eliminate the need for ten base-load nuclear power plants."

Because of this energy waste, and until efficiency is designed into them, the familiar refrigerator might become our best source of new space heating in the house. Someday, the refrigerator might be used both to chill food and warm the kitchen.

Another source of space heating, although not as unexpected, is the clothes dryer. Those who use the dryer in the wintertime sometimes disconnect the vent to the outside and redirect it to circulate the exhaust air indoors. Because there are problems associated with this form of direct venting, a new venting system called the Dryer Mate [6] has been created. What most people don't realize is that a clothes dryer sucks in air from the room which it then heats and mixes with the clothes to dry them. If the dryer draws in moist air released from the exhaust system a few moments before, then it will be working to dry the clothes with water-saturated air. On the other hand, if the dryer is used for an hour with the exhaust vented to the outside, it will suck most of the warm air in the house into the machine and then release it out of doors.

Therefore, what the Dryer Mate unit does is vent the exhaust air away from the dryer. A vertical pipe extended above the dryer contains a lint filter and a small fan which blows the air into another part of the room. The fan releases a welcome supply of moist air into the house and reduces the possibility that the air will immediately be drawn back into the machine.

Ironically, humidity in a room, as opposed to moisture on the skin surface, can increase the comfort level in a cool room. However, if the moisture in the air is too high, it can migrate through the outer walls of a room and condense against the insulation or the wood frame. This can result in mildew, the swelling of structural wood, or wood rot. Therefore, the best level of humidity is usually around 40 percent.

In addition to the Dryer Mate, there are a few other simple ways to introduce a small amount of humidity into a house. One way is by using a humidifier built into the heating system or placed near a return-air duct. Other methods include releasing the steam from a dishwasher into the room after the wash cycle is completed, washing dishes by hand (which also warms the hands), letting hot water from a shower or bath stand until it cools down, or by hanging clothes inside to dry.

An item whose value as a space heater is questionable, yet which does release a small amount of welcome moisture into the room is the English-designed, electrically heated towel stand [7]. Although an upright old-fashioned steam radiator or exposed hot water pipe will do the same thing if a towel is draped over it, this heavy duty, 3/4-inch, chrome-plated, oil-filled tubing (for even distribution of heat) will rack-dry the towels in style.

The use of oil-filled tubing to evenly distribute heat is a good idea. Oil is a better heat storage medium than water or air. It also can become hot when it is spun, shaken or run through an extended length of pipe. In fact, workers on the Alaskan pipeline worried about the friction-related heat problems that would be created by Alaskan oil running through the pipe over frozen tundra.

One man who discovered this phenomenon several years ago and used it to advantage was Eugene Frenette of Derry, New Hampshire. He patented a "revolutionary" electric space heater called the Frenette Furnace [8] which uses hydraulic oil inside a series of two drums. The drum-within-a-drum design uses friction generated by the small, inner, spinning drum to produce heat in the oil contained by the larger, outer drum.

The Frenette Furnace can be used as a forced air furnace and tied into a standard central heating system or it can be used as a space heater in the living room. While the space heater is only coming on the market at the time of this writing, plans for the first units call for decorator wood-grain panels to cover the heater. The space heater will then look like a two-foot tall, 18-inch wide coffee or end table. The Frenette Furnace can be plugged into any standard electric outlet, and because it does not burn any fuel it does not require an outside vent or chimney to exhaust fumes.

A similar use of friction for space heating has been talked about in connection with small windmills. The windmill rotates a paddle wheel resembling the blades of a washing machine inside an insulated chamber filled with a liquid such as water. The friction-generated heat is produced in a manner similar to that of the Frenette Furnace. If the heat is conducted from the insulated chamber into a storage reservoir such as rock storage, it can be released into the house when needed.

Although solar technology is making slow headway in the housing market, simple solar devices which can be placed on the outside of the house to draw in the sun's heat are now available. One of these is called the Mother Earth "Heat Grabber." It is a small wooden box that draws solar heated air through a window into the house but doesn't pull warm air out. The small box can heat a reasonably well insulated room on a sunny day, regardless of cold temperatures. It takes less than two hours to build and costs under $40.

The Heat Grabber [9] is simply a weathertight wooden box that's insulated on the bottom and topped with single-strength glass. Inside this box, an insulated divider is mounted just about halfway between the glass and the container's bottom, with a small airspace left at the foot of the box for circulation.

The divider itself is additionally separated into two parts: the main body (which serves as a partition between the upper and lower chambers of the collector) and the airflow passages (which are located on the top side of the divider under a black aluminum plate and which—in effect—act to increase the surface area of the collector).

The divider is then brought out the upper end of the box where it forms part of a "lip" that hooks over a windowsill (so the window can be pulled down snugly against the top of the collector's frame and the glass). . .leaving the main part of the collector leaning against the outside of the house at about a 45-degree angle." The box draws air from the house and heats it while the sun is shining and returns it to the room in cyclic fashion. When the sun isn't shining, the cool air sinks to the bottom of the box and prevents any air exchange from occurring.

Another interesting solar heater still under experimentation that has many solar designers anxiously awaiting its perfection is the "phase change" eutectic salt tile panel. Used as a demonstration model in a building at the Massachusetts Institute of Technology (MIT) and then installed under a contract from the Department of Energy on a 1200 square foot branch library in Mendon, Ohio (still under construction at the time of this writing) the eutectic salt tiles offer a simple heating system. They can store about 5 to 10 times as much heat as an equal volume of water can store, and have 15 to 30 times the storage capacity of an equal volume of stones.

When sodium sulfate decahydrate, known as Glauber's salt, is mixed with other chemicals and sealed in thin-walled shallow boxes or trays, it will both store and radiate solar heat. At the MIT demonstration project the eutectic salt is placed inside the flat, thin, polyethylene-lined aluminum foil bags and set inside the hollow core of ceiling tiles. Sunlight is then reflected onto the 2-foot-square tiles by slim shade reflectors mounted in the windows. Once set, the reflectors must be adjusted once a month to compensate for changes in the angle of sunlight.

Another way to use eutectic tiles for space heating is to place the tiles on a moveable stand such as a small table with rollers. The tray is set in front of a window during the day where it absorbs as much heat as possible. It can then be rolled to a favorite chair or into the dining area, for

example, to serve as a small space heater. This method of heating, however, is mentioned only for illustration and to this writer's knowledge has been used only for experimentation.

One final method worth mentioning because it may have future residential use is the so-called "Heat Mirror" [10]. The mirror is created when a factory-installed plastic film is stretched between two panes of glass. The mirror is transparent to short-wave (solar) radiation but not to long-wave (heat wave) radiation. This special film allows sunlight to enter a room then prevents the heat from escaping back outside through the glass. The heat loss is cut 75 per cent while the transparency of the window is decreased about 20 percent.

Finding ways to warm the body by using sunlight doesn't have to involve high technology and complex systems. Simple, direct exposure to the radiant heat of the sunlight will keep you warm. Likewise, if you placed an object between yourself and the sun, blocking out its light, you would feel an immediate drop in temperature, perhaps even discomfort.

The ancient Greek philosopher Diogenes (412-323 B.C.) once felt such a moment of discomfort. According to legend, Diogenes was stretched out naked beside the tub in front of his house, enjoying the sun's radiant warmth. Suddenly he realized that a small group of travelers was coming down the path toward the house. This unexpected company turned out to include the warrior Alexander the Great and his entourage. The great warrior marched up to the spot where the equally famous philosopher was now sitting. Casting a large shadow over him, Alexander proceeded to engage in conversation.

After a short time, the warrior found himself so inspired by both the dignity and the insights of Diogenes that he desired to offer him something in return. Alexander told the philosopher that he could ask any wish and it would be granted.

Without hesitation, the wise Diogenes gazed up into the face of the warrior who stood towering over him and said, "Then stand out of my sunlight!"

Startled, Alexander granted his request, but as he turned to leave, he told his followers, "Were I not Alexander the Great I would prefer to be Diogenes."

The next section describes some ways we can bask naked as Diogenes in the sunlight in the middle of winter without feeling the cold.

REFERENCES AND NOTES

1. Gustave Doré and Blanchard Jerrold, *London: A Pilgrimage* (New York: Dover Publications, Inc., 1970), p. 114.
2. An item found in the catalog of Hammacher-Schlemmer, 145 East 57th St., New York, N.Y. 10022.
3. Evan Powell, "Thermal Storage Furnace", *Popular Science Magazine*, November, 1977, p. 70.
4. Aztec Marketing Co., 11575 East 40th Ave., Denver, Co. 80239.
5. Hammacher-Schlemmer catalog.
6. Dryer Mate Co., Box 1366, Dayton, OH. 45401
7. Hammacher-Schlemmer catalog.
8. Frenette Friction Furnace, Box 255, Derry, N.H. 03038.
9. "Heat Grabber Is Back," *Mother Earth News*, November/ December 1978, p.95.
10. For more information on the Heat Mirror, a thin, plastic film that is stretched between two panes of glass and is transparent to short-wave but not long-wave (heat wavelength) radiation and thus keeps the heat inside the house, contact Suntex, Corte Madera, Ca 94623.

HOT TUBS

SAUNAS

"In the winter they often go out completely naked and roll themselves in the snow, while the temperature is 40 or 50 degrees below zero. They wander naked in the open air, talking to each other and even with a chance passerby. . . " Guiseppe Acerbi, as quoted in H.J. Viherjuuri, *Sauna: The Finnish Bath*

Sometimes, craving warmth in the winter, we construct images of warm places in our minds, mentally flying south for the season to recline in the sun on some tropical beach.

For the price of a ticket south we can get some of the pleasure of going there without even having to pack a bag. The more money invested, the more grandiose the plan. Unlike vacations that end up reduced to a 3 by 5 inch postcard pressed into a scrapbook, a hot tub, a sauna, a walk in a solar collecting greenhouse or a solarium can provide us with a place to get away from cold winter weather.

A hot tub or sauna can turn the body into a heat sponge in the wintertime. Similar to fueling a woodburning stove, the body will stoke up heat and radiate it for hours. By placing the hot tub in a greenhouse or solarium we can submerge in the warm water on a sunny winter's day, nose to marigolds, eyes to ripening tomatoes and enjoy the pleasures of a tropical resort right at home.

The concept of bathing for warmth is new to most of us. Unlike the Japanese who wash before they bathe, Americans use the bath for cleaning. But bathing for pleasure, therapy, camaraderie and warmth is still a daily custom in many countries. It was also a custom practiced by many ancient cultures including the Egyptians and Babylonians, the Greeks, Turks, Romans, Hawaiians, Finns, Japanese and of course some of our American Indian tribes.

The Ancient Romans, for example, built enormous monuments to celebrate the pleasures of the bath. In their enthusiasm, however, they nearly denuded the forests surrounding Rome to provide fuel for the furnaces which kept the public bath waters hot.

The public baths, or thermae, of Rome were of gigantic proportions. One, in fact, was able to accommodate 18,000 bathers. The tiled thermae were supplied with water from aqueducts. The water emptied into huge reservoirs and was heated by a system of furnaces before being distributed to the bathing chambers. These buildings were monuments to some of the finest art and architecture of the times. Some of the largest buildings included numerous small shops around the perimeter, as well as steam rooms, gymnasiums, massage rooms and even museums and libraries.

With the decline of the Roman Empire, bathing fell into disfavor for many centuries in Western cultures. But in 1664, the famous diarist Samuel Pepys took a dip in the hot waters at the English resort called Bath, came home to lie in bed sweating for an hour after which he rose and recorded the event.

The hot springs of Bath which were first discovered by the Romans and then developed as a resort in the middle of the first century A.D. brought one-half million gallons of hot water (120 degrees Fahrenheit) to the surface every day. The Romans built a lead-lined reservoir to contain the water. They let the excess hot water (which now heats the buildings of the city) run through a culvert to the river. By the time Pepys entered the baths, they were in the midst of their second Golden Age.

Because the baths were popular and often overcrowded, Pepys sought to experience the pleasure of the hot water while avoiding the crowds. One morning, Pepys and his party rose at four o'clock and were each carried down to the baths in a chair. "By and by," wrote Pepys in his diary, "though we desired to have done before company came, much company came; very fine ladies, and the manner pretty enough. . ."

In the early 1700s bathers of both sexes walked about the warm water of Bath together but there was no chance for swimming for each was "respectably covered in long, complete, flannel bathing suits. The ladies had little trays, fixed to the waist and floating in front of them, which carried their handkerchief, puff box and snuff box."

In American society the idea of the bath ritual has seemed to be get it over with as quickly as possible. The invention of the shower is a perfect example of the drive for efficiency at the expense of everything else. But the use of the hot tub for keeping warm in winter may turn that around. The closest modern example of hot tub use for warmth and relaxation is the Japanese bath. The use of the hot bath for absorbing body heat is traditional in Japan. The Japanese periodically immerse themselves in hot water then wrap in long, heat-trapping robes or kimonos. In some parts of Japan, hot baths are prepared twice weekly. The family enters in order of seniority after having first cleansed themselves with soap and water outside the tub.

While the Japanese seem to prefer a scalding 120 degrees Fahrenheit, water for the American hot tub can be thermostatically controlled to maintain a constant temperature of 106 degrees. A cover which floats on top of the water when the tub is not in use helps to hold the warmth inside.

At present the popularity of the American hot tub has overflowed California, where it seemed to gain its first visibility, infiltrated the Rocky Mountain region—noted as far north as Montana—and has been reported as far east as Freedom, Maine where a hot tub company advertises unique cypress, nature tubs for sale with special, energy-saving heating devices.

For those traditionalists who have a hard time crossing self-imposed barriers to accept the hot tub phenomenon in this country as anything more than an X-rated playtoy, it is probably useful to take another look from a perspective which might be more familiar to us. For example, think of the hot tub as a giant-sized hot water bottle with the top cut off and made of wood. But, you climb into it instead of lean against it. On a cold night in the middle of winter, its primary function, just as the hot water bottle, can be to help the body store up enough heat to last a few hours or perhaps all night.

Hot tubs can be made from redwood, oak, pine, cypress, and sometimes even plywood. Some are large size wine or water storage barrels with the top cut off. There are fiberglass hot tubs too, but as Leon Elder says in explaining his preference for wood in his book, *Hot Tubs,* "wood is more loving. . . ." Hot tubs require more water than the traditional bathtub: a five-foot tub filled with 3½ feet of water takes 515 gallons of water, but the water, if not used for cleaning, can be reused. A six-foot tub takes 742 gallons of water and 13 people to fill it. Hot tubs can also be sized to fit one or two people.

In 19th century homes, before the days when one could simply turn on the faucet marked "H" and receive a steady supply of hot water, one had to either be content with a cold bath—reputed to be good for the complexion—

or else wedge into modest hip baths or shower baths. Those who could afford servants to carry buckets of hot water from the kitchen stove to the tub could lounge in the luxury of the hot bath without having to carry the water themselves. In addition to hand-carrying hot water, there were other ways of providing hot water for the tub. Benjamin W. Maughan invented the geyser, which stored water for the bath in a tank above the tub. As the water dripped down thin wires toward a gas flame, it would be heated to near boiling. One drawback to this method was that the gas flame had no flue-pipe and so released dangerous fumes into the room when the water was being heated for use. Another method, used in France, heated the water after an hour or two but also released dangerous fumes. A charcoal fire was lit in a deep pot which had two tubes emerging from the top: one to draw in fresh air and the other to release smoke. The English preferred to treat the tub like a kettle by placing a gas burner under it. The burner could be swung out to be lit and then returned beneath the tub to heat the water [1]. While these methods may cause some smiles, they are not without their modern counterparts. One tub, the Cannibal Pot, described by Leon Elder in *Hot Tubs*, resembles a giant concrete soup bowl, stuccoed on the outside and ringed by inlaid tile. Underneath the pot is a thick iron plate heated by a gas burner set below the plate. In addition, the U.S. army has invented an immersible kerosene heater, somewhat like the antique French water heater, which is used to keep garbage cans of dishwater hot.

Hot tubs can be heated by a woodburning fire, by more conventional gas, oil, or electric systems, or by a solar hot water system. For more detailed information on specific systems, you might want to pick up a book on hot tubs at the library or bookstore or consult a hot tub dealer.

The ideal temperature in a hot tub is 106 degrees Fahrenheit. The standard temperature in a sauna ranges between 190 to 200 degrees Fahrenheit. Although the boiling point of water is around 212 degrees, the body can take far more heat in a dry environment than in a wet one.

The sauna, like the hot tub offers a quick thaw to a winter-chilled body. After a sweat, the body takes a long time to cool down again. It can take stark temperature contrasts, from 200 degrees to below zero without any ill effects. But to take full advantage of the sauna's warmth in winter, it's a good idea to move to the bed quickly with only a quick rinse-off in the shower.

The sauna is most typically thought of as a custom belonging to the Finnish people. Indeed, it has been part of the ritual of the Finns for over 1000 years. But saunas or sweat baths have also been part of the communal ritual of other societies. The American Indian tribes had their sweat houses. The Romans included steam rooms in their multi-chambered thermae. The Turkish were accustomed to a more humid steam bath, called the "Turkish Bath" in this country. And the Russians, like the Finns, preferred the dry bath.

In man's early history, fire was often considered a gift of the gods. Some people have theorized, therefore, that the fireplace, and later, woodburning stoves, might have been used in early saunas as altars for worship. Stones piled high on top of the heat source might have been considered an altar both to offer thanks for the fire and to rid the body of evil spirits. Although the first sauna was built in the United States in 1638, the notion that it had a purpose other than relaxing and warming the body persisted in the minds of those who were suspicious of its value. In the 1800s, farmers in Minnesota, believing that the Finnish neighbors were "worshipping pagan gods in strange log temples and cavorting naked in the moonlight in what seemed to be ritualistic dances," brought the Finns into court for these so-called "strange" practices.

A hundred years earlier, the Italian traveler Giuseppe Acerbi described the Finnish sauna after having seen and experienced it firsthand. "Among the many peculiar experiences that I had in Finland, one was the sauna. Most of the farming population have a small building erected specially for bathing. It consists of only one small room, at the back of which a heap of stones is piled. These are heated with fire until they are red hot; then water is thrown over them forming a thick cloud as it evaporates. . . . All the time they are in the sauna, the Finns keep beating their bodies all over with young birch branches. After ten

minutes they get so red that they look frightful. In winter they often go out completely naked and roll themselves in the snow, while the temperature is 40 or 50 degrees below zero. They wander naked in the open air, talking to each other and even with a chance passerby.

"If a traveler in search of help happens to arrive in a remote village at the time when all the inhabitants are in the sauna, they will leave the bathhouse in order to harness or unharness a horse, to fetch hay, or do anything else without ever thinking of putting any clothes on. Meanwhile the traveler, although enveloped in a fur coat, is stiff with cold, and does not dare to take off even his gloves, however used he might be to exposing his extremities to the air in other parts of the world.

"Thus, the Finns move in less than a few seconds from a heat of 170 degrees to a cold of minus 50 degrees, which makes a difference of more than 220 degrees, and the effect is very much the same as if they jumped from boiling into freezing water. What astonishes the people of our climate most is that no ill effects ensue from this sudden change of temperature. People who live in warmer climates, on the other hand, are extremely sensitive to a change of as little as ten degrees, and are liable to get rheumatism even when the most gentle wind blows" [2].

The Finns originated the concept of the sauna to provide a source of warmth and comfort to offset the harsh living conditions of their everyday lives. Today there are over a million saunas in Finland. In the United States, where their popularity began to increase in the 1960s, there are now thousands of privately owned saunas.

Public saunas and neighborhood saunas, popular in Finland, have not caught on as quickly. However, "time-shared" saunas, a means of sharing warmth as well as expenses, can be built by several families living in the same building or on the same block. Each family can use the sauna according to a prearranged schedule.

Saunas do not have to be housed in separate buildings. In a private home, "retrofitting" can mean that the bathroom might also double as a sauna. In one sauna, built in a 6- by 11-foot bathroom, the bathtub is used for an after-sauna plunge. Two-tier redwood benches double as towel storage shelves. The "hot rocks" sauna unit takes about 30 minutes to reach 180 degrees Farenheit. And as with the hot tub, several people can take advantage of its warmth at the same time.

A word about greenhouses. According to conventional wisdom, a greenhouse is a building that artificially recreates the heat and light of summer—indoors—to force the growth of plants or extend their growing season. But a greenhouse can also unite plants with humans for mutual heating benefits.

For those who look at green plants as root and stem pipelines to a solar heating system, a greenhouse can become (for both heating and income tax purposes) a walk-in thermal storage unit or an attached solar collector.

There are many greenhouse designs and many reference books for building a greenhouse. However, the greenhouse that is built on a south or southeast wall and stores and circulates surplus heat for use in the house will keep you warm in winter. Costs may run anywhere from $200 to

$30,000, at 1979 prices. Sunlight can be drawn in through glass or plastic-covered walls and stored in a thermal mass for later use. The thermal mass can include water or rock storage systems, a darkly painted masonry wall, or brick or flagstone tiles on the floor. These or similar materials will store the accumulated heat. In one greenhouse, 55-gallon oil drums, painted black and filled with water, serve a dual purpose as a heat storage medium and as the foundation for the plant storage shelves.

A hot tub in the greenhouse can add humidity and also allow tub users to benefit from the pleasant surroundings. Humidity can also be introduced through other means: by growing plants in hydroponic systems or by indoor fish tanks (growing edible fish indoors).

To turn the simple greenhouse into a sun sponge, the heat that is trapped and stored inside must be kept inside. As with insulated or storm windows in the central house, the windows in the greenhouse or sunroom-greenhouse (called a solarium) must keep the heat from escaping through the windows at night. The windows can be temporarily sealed with insulating panels set in place for the night, and removed in the morning, such as sheets of thermax cut to size, or by double-paned glass, plastic sheeting or plastic paneling such as Kalwall panels. A one-half inch or greater layer of still airspace trapped between the two pieces of material will serve as fine insulation.

Simple, direct venting into the house, by opening an interior door or interior window, or by using rotating fans or more sophisticated venting systems, will move the warm air from the greenhouse into the mainhouse or into another temporary storage system, such as rock storage, for added warmth when needed.

REFERENCES AND NOTES

1. David de Haan, *Antique Household Gadgets and Appliances* (p. 147) Blandford Press, U.K./Barron's, U.S.A.
2. From *Sauna: The Finnish Bath* by H.J. Viherjuuri. Copyright ©1965, 1972 by The Stephen Greene Press. Reproduced by permission of the publisher.

BODY

BASICS

"If you are newly married people, go to bed. You need neither electric heat, nor light."
(Advice of the Swedish State Power Board during the Swedish Energy Crisis 1969-1970.)

The body is a space heater as well as a heat receiver. To keep warm, the body must maintain a heat balance: it must receive as much heat as it radiates.

The sources on which the body depends for warmth transfer their heat to the body in one of three traditional ways: by conduction, convection, or radiation. A fourth method is through the use of biofeedback wherein the power of the mind is used to regulate the blood flow in order to elevate the body temperature.

Heating by conduction involves a direct transfer or warmth. A hot mug of coffee, for example, *conducts* its heat directly through the container to the cooler hands which cradle it. An electric heating pad or footwarmer conducts its heat the same way.

A central heating furnace conveys heat indirectly, warming air to a desired temperature and then bringing it into a room. The warm air mixes with or replaces. the cooler air surrounding the body.

Sources of heat can also warm the body directly through radiation. A woodstove *radiates* heat to the person sitting before it, while an object placed behind it will remain unaffected by the stove's warmth. The sun radiates heat the same way. It does not warm the air between itself and the earth, but warms objects directly, according to their ability to absorb heat.

While the body readily absorbs heat from the sun and other sources, it is also a source of heat. Clothed or unclothed, the body continuously radiates heat to the air and to the surface of objects surrounding it. The body constantly manufactures and releases moisture and heat as by-products of body metabolism and physical activity. Thus the old saying: Wood heats three times: once in the sawing, once in the stacking, and once again when it is burned.

The amount of heat the body radiates can be measured in Btus. Even better, the amount of heat produced can be measured by equating it with familiar items:

A single body at rest radiates 350 Btus an hour, or the heat-generating equivalent of a 100-watt bulb.

Three people sitting generate the heat equivalent of a small electric space heater.

Three people with a distraction generate the heat of a large electric space heater, which is also equal to the amount of heat generated by a person engaged in heavy manual labor or in an activity such as skiing or jogging.

The heat radiated by the bodies of twelve people in a room is the space-heating equivalent (. . .the poor man's equivalent) of a small woodburning Jotul stove.

The value of the body as a heat source for buildings is only now being *re*discovered by high technology societies. A few experimental buildings have begun to utilize the heat generated by the human body; for example, Canada's two-story Saskatchewan Conservation House in Regina is so well-insulated that according to its designer, it needs "only a single light bulb and two couples making love" to heat it. Actually, the house can be heated to a comfortable temperature with an electric heater about the size of a rolled-up newspaper.

Ontario Hydro's new 20-story office building in Toronto uses the body heat from its 5,000 employees, plus the heat from the electric lights and office machines to cut the heating costs in half.

Fans located on each floor draw the warm air out of the partitioned offices and down into the basement. Here the exhaust air is circulated through a giant compartmented water reservoir which in turn releases heat into the building. A simple form of cogeneration. . . .

The following pages describe some other ways for making good use of body heat to keep you warm, as well as what to gather around you to stay warm.

"Why not go ahead and wrap yourself in insulation to keep warm, you've already got a pot-belly!" Mathew Morrison

Living without central heat, one learns the art of the reverse strip tease. Because a person's comfort depends upon what is happening on the surface of the skin, clothing must serve as body insulation. To provide warmth, clothing must trap a layer of still, warm air against the skin, draw away any moisture, and perform both of these functions at the same time whether the body is active or at rest.

How we clothe ourselves "in society" has typically been a function of style. We dress to imply social status or social attitude. But there was a time—prior to the introduction of the central heating system—when the style of clothing also depended upon its ability to provide thermal comfort.

Just as our houses were once built as simple shelters designed to keep out the cold, our clothing also had to keep us warm. We could not simply turn up the thermostat and warm all the air in the house to a degree we thought comfortable, for there was no thermostat and no central heating system. Instead we could add another layer of clothing.

Up until the mid-20th century, all clothing was made of natural fibers of wool, linen, silk and cotton. These fabrics trapped the body heat naturally and also drew moisture away from the skin. By wearing several layers of clothing indoors as well as outdoors, a person increased the insulation value created by the trapped warm air. In the wintertime, the clothing was not lightweight and skimpy but heavy and durable, and it covered the entire body.

A floor-length cloak with an attached hood (to protect the neck from drafts) was in fashion for centuries. For some people, the cloak could even be used as a blanket for their beds. Cloaks worn by knights in Europe sometimes left only an opening for the eyes. When cloaks were turned around by children, they could be used in a game of blindman's bluff. And, of course, Little Red Riding Hood wore one to her grandmother's house.

Hats were also considered essential cold-weather clothing. Who ever heard, for instance, of a man entering or leaving a house prior to the 1700s without a hat on his head? A cap was worn both for comfort and sometimes to comply with a law. In 1571, England passed a statute requiring all common men to wear a wool cap on Sundays and Holydays. Even King Edward VI wore one. The statute was intended to encourage the wool worker's pride in their country. But the wool cap also kept the head warm. When wigs became fashionable in the early 18th century, the wearing of a hat indoors became a distinct sign of low breeding.

These wigs were hot and heavy, but whenever they were taken off, a cap temporarily took their place. Caps, in fact, were also worn to bed. They were a standard part of the bedclothing of men, women and even children until modern times.

In Colonial America, while the settlers attempted to keep up with the changing fashions in Europe, their winter clothing was styled primarily from wool, leather, canvas and other "strong, durable stuff." And while Puritan clothing was criticized for being dull and drab, this drabness was, in reality, the result of their preference for the warm tones of brown, which varied even to include orange, and for the "sad-colors" of purple, deer color, and French green.

Women in Colonial times wore long, floor-length skirts. In the wintertime, some even wore several layers of skirts. In time, well-to-do women began to wear layers of plain-colored or blue and white striped wool and linen (linsey-woolsey) petticoats beneath an overskirt, while the overskirt was fashionably styled. Because the fashionable skirt was expensive to make, women of poorer means usually wore only the petticoats. A runaway female slave might even be described by the fabric of her skirt. However, for both poor and rich alike, the floor-length skirt trapped body heat between the layers and folds of the cloth. It also kept women's legs free from drafts. An exception was the hoop skirt, which lost its circle of warm air whenever it was bumped or tipped up.

To protect their arms from the cold weather, women kept them covered with long sleeves, often puffed or ballooned (to trap more body heat) and tight at the wrist (to keep heat from escaping). A shawl kept the back free from drafts while it left the front open to the radiant warmth of the woodburning stove or fireplace.

Men wore vests, jackets and cloaks indoors as well as out. And in the early 19th century, despite an initial outcry of opposition, trousers becams a fashionable item of clothing. While the Ancient Briton men and the Danes had worn a form of trousers, long pants had previously been worn only by men of the peasant class. The introduction of long pants into all classes of society opened the way for the invention of a special kind of long underwear: the union suit. This long, woolen undersuit provided equal distribution of warm clothing over the entire body.

The popular union suit was also a byproduct of the Dress Reform Movement in Germany in the 1870s. This movement, according to Alice Morse Earle's delightful history of clothing, *Two Centuries of Costume in America,* was based on the assumption that wool was "the only proper covering for the human body" [1]. For a time everything was made of wool, from handkerchiefs to underwear.

By the late 1800s central heating systems began to replace the fireplace and woodburning stove as the main source of home heating. With the advent of this steady, inexpensive supply of warm air to every room, people could easily turn up the thermostat to adjust to cooler temperatures instead of adding an extra layer of clothing.

In such a controlled environment, both men and women could wear lighter-weight clothing throughout the year. Women's skirts could be elevated to the knee and higher, if desired. Over the years central heating has created a paradox; by turning up the thermostat, we can wear lightweight clothing indoors all winter long; in the summertime we often need a sweater in air-conditioned buildings.

The concept of clothing as insulation was not seriously studied until the 1940s. At that time, the military became concerned about providing adequate protection for soldiers on the battlefield. Physical conditioning and training, their studies had shown, would not increase a soldier's acclimation to the cold. A soldier suffering from exposure to the cold could not be expected to fight a war against the enemy successfully.

To restrict the loss of body heat and maintain the soldier's comfort on the battlefield, it therefore became necessary to determine what type of clothing functioned best as body insulation.

Overclothed, a soldier would be prone to excessive sweating. Moisture trapped in the fabric or on the skin accelerated the flow of heat from the body. It also accelerated symptoms that might ultimately lead to a drop in deep-body temperature, or hypothermia, and possibly to death. Underclothed, the soldier could lapse into uncontrolled shivering which would likewise accelerate body heat loss and lead to the same unhappy end.

Therefore, properly clothing the soldier to enable him to function as he was intended became a military science that only recently has found application in the civilian sector.

For example, for years the soldier's oversized black leather boots were laughingly called *Mickey Mouse* boots because they looked like the oversized shoes worn by the famous Disney cartoon character. But they had been designed that way for a reason.

When soldiers returned from duty in Alaska during the early 1970s, many sold the boots off their feet to Alaskans living year round "in the bush" and to construction workers on the Alaskan pipeline for $200 to $300 a pair. The boots kept the feet warmer than any others, and for those who bought them, warmth with or without style was worth the price they had to pay.

Though clumsy looking, the "fat" boots like today's "Moon" boots added an extra layer of air around the foot. Increasing the still-air space around the foot added more warmth than an extra pair of wool socks. In fact, extra socks in ordinary boots might even make the feet colder. The socks substituted fiber, a conductor of heat, for air, which served as insulation. And if the socks were a tight fit, they constricted the flow of blood to the feet.

Although fiber is a conductor of heat, cloth made from natural fibers can also contain air within the fabric, slowing the heat loss.

Until the 20th century, all clothing was made from natural fibers, which included wool, cotton, linen and silk. But by the late 1950s, synthetic fibers began to dominate the clothing market. While the use of cotton and wool remained about the same, the use of synthetic fibers more than quadrupled between 1960 and 1973.

Synthetics offered special advantages that made them desirable. Coming on the market when the price of home heating fuel was lowest and when clothing was no longer needed to protect the wearer from cool indoor temperatures, the petrochemical materials—nylons, acrylics and polyesters—quickly made up most of the ready-to-wear market. Lightweight, wrinkle-resistant as well as wind and water repellent, they offered a convenience that could not be matched by natural fibers.

However, they also created new problems for those who purchased them. Clothing made of pure cotton felt cool in summer and warm in winter. But synthetic materials created problems of increased warmth during the summer and excessive coolness during the winter.

Those who took up the science of clothing discovered that body moisture becomes bonded to the synthetic material, increasing the feeling of wetness or clamminess close to the skin. The moisture is present because the body is six percent sweat-wetted at all times. Increased physical activity accelerates the rate at which sweat is produced.

Because moisture increases the conduction of heat from the body surface, synthetic materials actually emphasize the loss of heat from the body in cool weather. In addition, synthetic fabrics contain larger pores within the weave than natural fibers and lack the "nap" or webbing of natural fibers which covers the pores and traps rising warm air. Without this nap, body heat passes freely out of the synthetic clothing.

In the late 1970s, people began to reevaluate the use of synthetic materials. As the high price of heating fuel drove down the average temperature in most American homes, clothing made from natural fibers, along with down-filled clothes, began a resurgence in the market. To compete, the manufacturers of synthetic materials had to find a way to technically duplicate the functions that cotton, wool, silk and linen did naturally.

They found a way to bond cotton and nylon fabrics to the inside of a windproof material. This allows the clothing to hold the moisture away from the skin, while at the same time, the outer layer holds the body heat in.

Sports equipment manufacturers discovered Gore-Tex, a

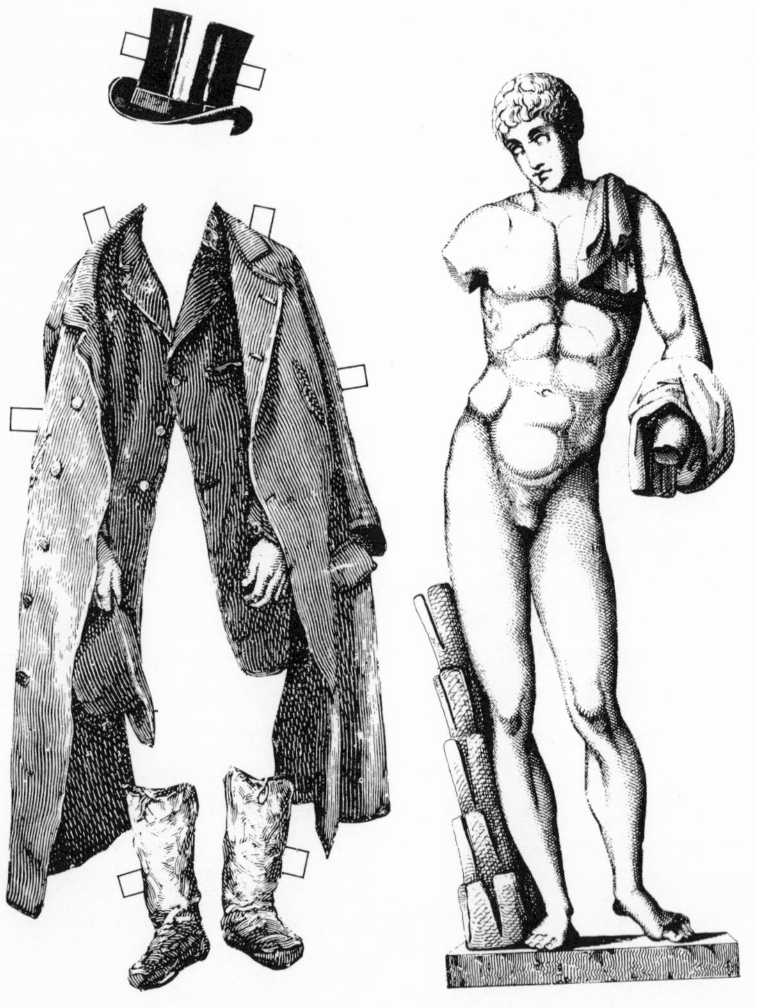

synthetic material that "breathes" like natural fibers but is wind- and water-proof. The thin, teflon-like plastic material contains nine billion pores per square inch, pores large enough to allow water vapor to escape from the skin but small enough to keep rain-drops out.

In recent years, clothing styles have begun to include the use of reflective materials. The "Apollo suit," for example, is made of aluminized, vinyl-like material bonded to nylon. The reflective materials greatly increase the amount of heat retained by the body with little increase in fabric thickness or weight.

The problem with reflective materials, however, is the same as with all synthetic materials. They trap the body's heat and create a moisture problem, because, as mentioned, even without exercise our bodies produce moisture on the surface of the skin.

With natural fibers, moisture goes into the clothing when you sweat, and is removed from the skin's surface. Because moisture, as well as fabric, is an excellent conductor, it draws off the body's heat if left in direct contact with the skin. But typically, there is a thin layer of insulating air between the skin and fabric which moderates heat loss and maintains its own temperature. This insulating layer can be adjusted by adding or subtracting layers of clothing.

With reflective materials, moisture evaporates from the skin's surface, rises to the inner surface of the reflective fabric, condenses and returns again to the skin. This recycled moisture only increases the loss of heat from the body.

To correct this problem, some clothing manufacturers are turning to synthetics such as Lifa Super, a product of Norway, presently used for long under-wear. It acts as a fine wick to draw moisture into the outer layers of clothing while remaining dry against the skin. Others are working on manufacturing techniques to blend natural fibers with reflective foils so they become part of the insulating material.

In some materials, reflective layers are spaced to create a chimney effect, encouraging the necessary ventilation of body heat. Some impermeable clothing is designed with shoulder yoke vents. The "pumped air" effect of midlength or loosely fitting, poncho-style outergarments also helps ventilate the body and prevent moisture from condensing on the skin.

Of course, one extreme solution to the heat versus moisture problem is to design a totally controlled clothing system. Military clothing specialists have considered this as a possible means of dealing with extremes of cold such as those found in the Artic. Their encapsulated suit is similar to the suit an astronaut wears for space travel, but it is designed to deal strictly with the cold. The system provides a Buck Rogers style forced ventilation system that blows ambient air across the skin.

Until this system is perfected, the the military seems content to stick with one they know works quite well: the seven-layered Arctic Suit, designed for temperatures well below zero. The seven layers include: long underwear, a trouser liner, trousers, an outer trouser lining, second outer trousers, a parka lining, and a long parka made of tightly woven material [2].

However, the military's twenty-eight pound Arctic outfit is probably not a suitable model for us to follow in trying to keep warm. In most cases, we are simply adjusting to a drop of the thermostat to 65 or 55 degrees Fahrenheit, not minus 55 degrees. Still people suffering from cold hands and feet or contemplating a winter with little or no heat may consider their situation just as extreme.

Since a person's comfort depends on what is happening on the surface of the skin, the body's sensitivity to cooler temperatures can be lessened by adding layers of clothing. Thick, fluffy fabrics should be included among those layers. They are warmer than tightly knit or thin fabrics because additional body heat becomes trapped in the thicker fabric.

Tightly woven cotton trousers can be replaced with lightweight wool ones, loose enough to permit the addition of long underwear if needed. This combination of wool trousers and long underwear allows a four-degree drop in air temperature without a loss of comfort.

A blouse or shirt can be covered by a light or heavy-weight sweater or vest. (The English use an undervest which used to be called a "liberty" suit). Adding a light sweater permits the thermostat to be lowered an additional two degrees without a change in body warmth. A heavy sweater permits a drop of four degrees. A jacket loose enough to cover all the rest can be added last.

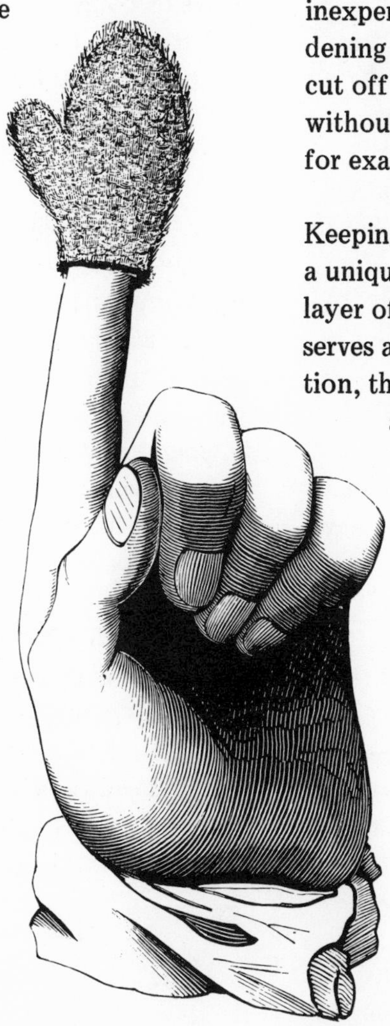

Out-of-doors, mittens provide better warmth for the hands than gloves which separate one finger from another and increase the amount of body surface exposed to the cold air. But mittens or gloves can also be worn indoors. If your work or leisure activities involve the use of your hands, you can wear an inexpensive pair of cotton gardening gloves with the finger tips cut off to keep your hands warm without interfering with typing, for example, or fine needlework.

Keeping the head warm presents a unique problem. A ½-inch layer of still air around the body serves as an insulator. With motion, the warm-air layer is drawn away from the body and the exposed skin feels cool. When the air temperature is significantly lower than the skin temperature, the body's automatic response is to condense to prevent the loss of deep-body heat. Blood vessels tighten, slowing heat loss through skin pores.

However, the head area cannot constrict. While the rest of the body closes down in response to colder temperatures, the head remains, figuratively speaking, like an open conduit. Body heat passes freely from the top of the head. Thus, old-timers and those who study human comfort will both advise: to keep the body warm—to warm your hands and feet—put a hat on your head.

Very young children and very elderly people pose special problems in cold weather. Children, for example, don't have to lose as much as heat as adults to feel as cold, because they have not built up layers of body fat and because their central nervous systems, which regulate the amount of blood flowing at the skin's surface, are not fully mature. Likewise, a low thyroid level or hypothyroid condition can also make an adult feel cold. The hormone liberated by the thyroid gland oversees general body metabolism. Fewer hormones reduce the body metabolism and lower the body's energy reserves, including the levels of stored fat.

The elderly produce less heat because their metabolism rates fall to as much as 15 to 20 percent less than when they were 20 years old. As the body grows old older, body functions become less efficient. Lacking the reserves of younger adults, the elderly become less active because it is harder for them to do the same physical work, and therefore they generate less heat.

Since both the young and elderly may suffer from the cold or become hypothermic before the average adult begins to feel any discomfort, special attention must be paid to them in the wintertime. If the surface of a very young or elderly person's skin feels cold, he or she must be provided with extra warmth. Extra layers of clothing or bedding (an electric blanket, underblanket[3], or *Thermos* emergency blanket) wrapped around the shoulders of an elderly person will trap additional heat and hold it close to the body. A lap robe, or a baby blanket used as a lap robe, will also provide extra warmth and comfort. However, if too many layers are added, causing the body to sweat, noticeably, then a layer or two should be removed.

REFERENCES AND NOTES

1. Alice Morse Earle, *Two Centuries of Costume in America 1620-1820* (Rutland, Vt: Charles E. Tuttle Co., 1971. Originally published by Macmillan Co., 1903)

2. A special thanks to Dr. Ralph Goldman, Director of Military Ergonomics Division, Department of the Army, Natick, MA for the time spent teaching me the fundamentals of a class he teaches on clothing and human comfort at MIT.
3. A thick blanket sold in some parts of Europe and the United States. The blanket has the texture of lamb's wool and is placed beneath the sleeper.

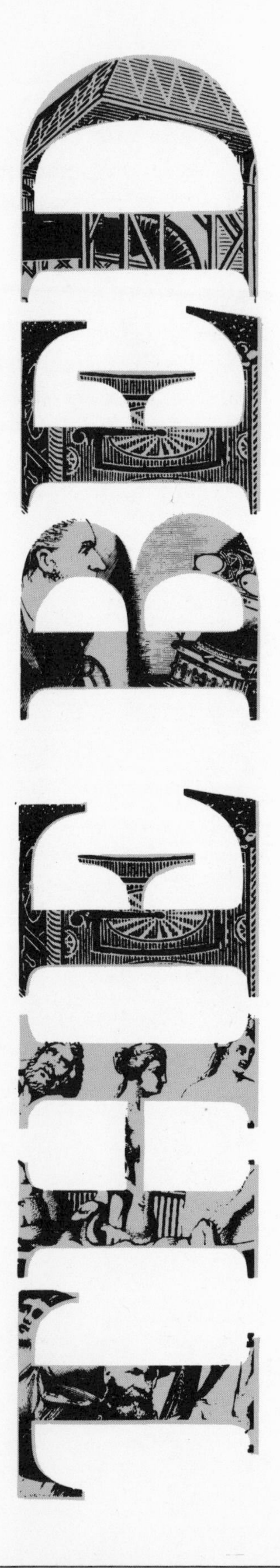

"When I was in college, I once went home with a friend on a winter break. They didn't heat the bedrooms, so we slept in the same bed. No one thought a thing about it. Nowdays, of course, people would look at it differently: two fellows sleeping in the same bed. . . ." Mad River Jack

Our clothing should keep us warm in the daytime. Our beds—where we spend nearly one third of our lives, the coldest part of each 24-hour day—should keep us warm at night.

To maintain a pool of warmth in a cold room, to make the bed a welcome source of comfort, three essential elements must be considered. They include the design of the bed, the type of bedding "insulation" to be used, and the source or sources of heat that can be added.

If we think of the body as a space heater, it is easy to understand why the kings and noblemen who lived in drafty castles in Europe often chose to make their home for the night in a niche in the wall with a heavy

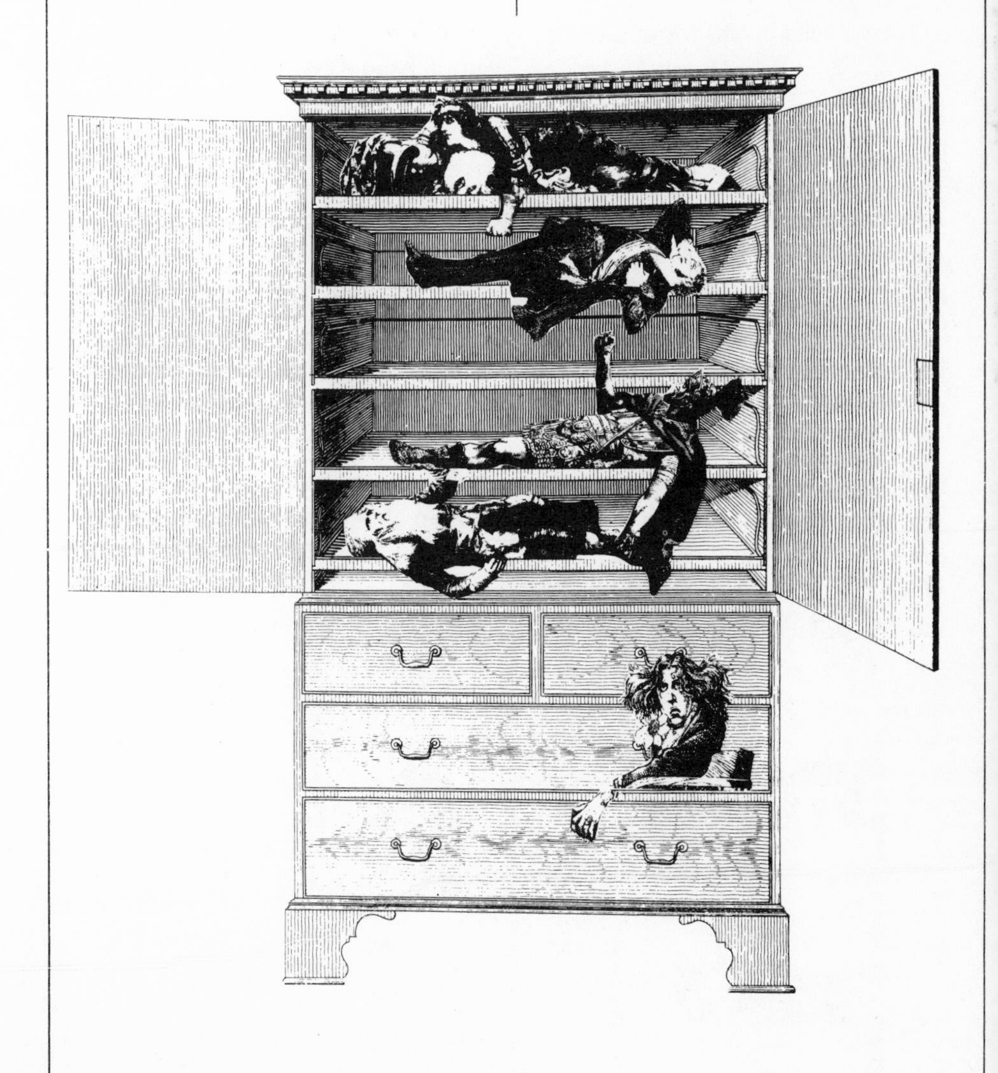

curtain pulled across the alcove opening, their body heat and a couple of blankets could keep them warm regardless of freezing temperatures in the drafty room.

Thomas Jefferson created such a niche for himself at his home in Monticello. But because he built it into a wall that was only as wide as the bed, curtains had to be pulled on both sides.

To save space in the room and to trap the heat, the Dutch built their beds into cupboards.

In 17th and 18th century America, the term "bed" was meant to include the bedstead, mattress and hangings altogether. The hanging for the four-poster or "tester" beds usually consisted of a top, a cloth draped at the head, a curtain at each post and a top and bottom valance. The bottom valance formed a draft barrier around the bed. When the curtains were drawn on all four sides, only 18 cubic feet of airspace had to be heated by the sleeper.

The curtains used in the winter were made of heavily woven textiles. They were taken off and aired from time to time, especially if there had been an illness in the house. In the summertime these heavy curtains were exchanged for lighter, more airy ones.

Some beds also doubled as couches in the daytime. But unlike the four-poster styles, their heavy curtains were pulled along runners or tracks attached to the ceiling.

This multipurpose bed is still in use in present-day China. The curtain falls from a central "crown" hung from the ceiling over the bed and is bunched up and pulled to one side during the day to allow for other uses of the room. A modern equivalent of the canopied bed might include roman shades, woven wood or bamboo curtains.

In addition to surrounding a bed with cloth or wood, you can build a warmer bed by placing it where the heat is: up high, loft-style, or up over a stove. An innovative 19th-century Englishman carried this idea a step further. He brought the heat directly to the bed by a fluepipe connected to a small lamp stove. The idea, patented in 1859, was to circulate the heat throughout the bed in pipes which would also make up the structure of the bed.

The Koreans were already using this idea, but had built their heat-carrying fluepipes directly into the floors of their houses. The Korean floor was described by the famous architect Frank Lloyd Wright, who adapted the concept to warm the floor of his Usonia No. 1 house. As he described it, the floor "was part of the chimney and thus extracted all the heat from the smoke that would have gone outdoors."

The Korean floor was made of adobe-like material with the heat corridors formed beneath the floor surface. When a fire was built in an oven placed at ground level, below the raised platform of the house, or was released from a pipe connected to the cookstove, the heat traveled down the adobe corridors and warmed the thermal mass of the floor. Like a ceramic tile stove, the floor absorbed and released the heat slowly.

The tradition of sleeping on a warm floor allowed the Koreans to readily adopt the electric blanket when it first came on the market. Travelers to Korea during the 1940s reported finding the Koreans using electric blankets as sleeping mats. The blanket eliminated the need to warm the entire floor but sleeping on it caused the electric wires to deteriorate.

Today an electric blanket can be used with other blankets. If turned on a few minutes before sleep, it offers instant warmth in a cold room. A heated waterbed can also provide the same instant comfort. But the electric blanket and heated waterbed would be of little use in a power outage.

A description of materials that filled the lining of mattresses during the 1800s reads like a materials list for insulation that filled the walls of some of the Colonial American homes. Any lumpiness, or lack of comfort, however, was probably made up for by the warmth provided by the fill materials.

While some of the so-called "best beds" were those filled with thick hair, beds were also filled with dried cornhusks or alternating layers of moss, cotton, feathers or straw. The thick featherbed trapped the heat and rose to contour itself around the body. In the morning, a few swift shakes would settle the feathers back in place and smooth out the bed.

Continental-style, down comforters, new to the American market, now offer this same kind of nighttime warmth, but are meant to be pulled over the top of the sleeper as both sheet and blanket, and not to be slept upon. A down-filled sleeping bag, of course, will also double as a bed covering when opened out.

Quilted bed coverings are now enjoying a revival. However, years ago, quilts were more than mere coverings. They conveyed not only a sense of warmth, but a spirit of love, family, friendship and community. Quilts were given away to young people when they got married, to the family of a newborn child, to people who had lost their possessions for one reason or another, or as gifts of friendship.

Before the days of synthetic fill materials, which work quite well, the quilt was comprised of a bottom and a top lining filled with either cotton stuffing, down, or wool, and then sewn closed. The top of the quilt, sewn over the top lining cover, was a series of colorful squares of either plain or printed fabric. The squares were of uniform size and joined side-by-side or separated by cloth strips which also served as a border around the squares.

Quilts were pieced together at social gatherings called "quilting bees." Sometimes only the women got together, but other times the whole family came along. The work of quilting was carried on despite the surrounding revelry. Patterns for arranging the squares were traded freely and unusual ones were especially sought after.

Quilts and heavy blankets were not used only on the bed. They could be taken off and used as lap or shoulder robes (somewhat like today's quilted comfort sacks) to provide extra warmth elsewhere in the house.

A buffalo-skin robe, for a time, was also a welcome covering. Another type of covering that everyone liked was the felt blanket. Woodsmen had discovered that the heavy blanket used on paper-making machines, when cut into bedsize coverings, kept them quite warm. The blanket was called a paper-mill felt. It gathered the fibers of the wet wood pulp in the process of manufacturing paper, but in time it lost its efficiency and was taken off the machine. The heavy blanket could then be cut into any size needed. For a number of years, felt blankets were a standard item in many backwoods bunkhouses. Today, felt is made from wool that is matted and pressed. It is thicker than flannel but does not have a nap.

However, one thick blanket does not make a bed warm, unless that single blanket is filled with down or down-like filler. To trap warm air, bed coverings should

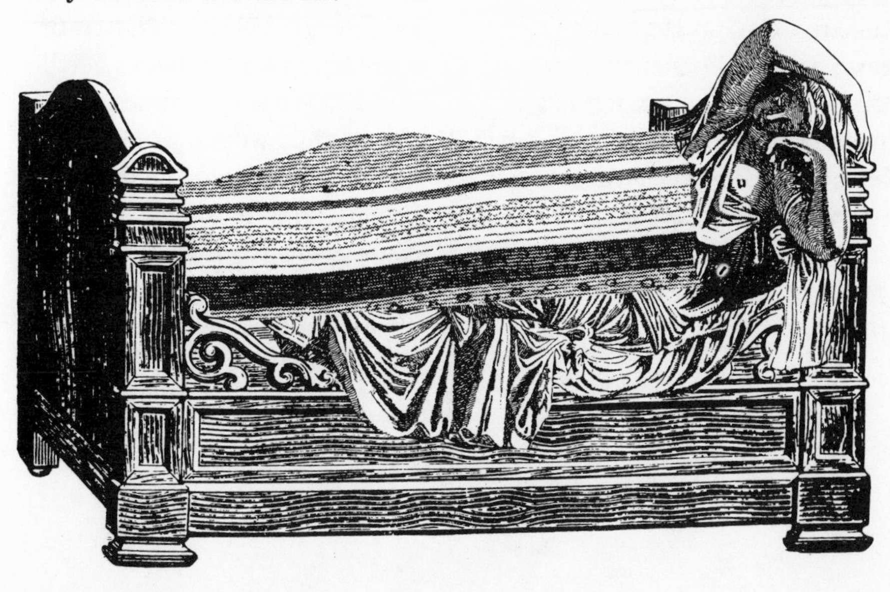

consist of several layers to trap warm air. Like a stack of pancakes, warm air becomes the "syrup" between each blanket as well as being absorbed and held in the fabric itself.

Sometimes, on very cold nights, a thin aluminum mylar sheet such as the Thermos emergency blanket (which costs only a few dollars at a sporting goods store) can be slipped between the blankets. However, because the reflective metal sheet returns body heat so well, it may make the body sweat too much. As with reflective clothing, this moisture might ultimately make the body feel cool rather than warm. Of course, one can simply remove the reflective blanket if it begins to feel uncomfortably warm.

Another blanket, popular in Europe, is the underblanket. The underblanket, as the name implies, is placed under the body but on top of the mattress. It is composed of two layers: a foam backing which keeps it from slipping off the bed, and an imitation sheepskin top comprised of tightly bunched coils of "hair."

The thick upper layer traps body heat in the open cells, thus providing a layer of warm air beneath the sleeper, while the fabric draws moisture away from the skin. Heat generated in this manner is comparable to that provided by an electric blanket. Cotton flannel sheets, though less effective, will also help.

Another method of trapping body heat is to slip a small ¼-inch or ½-inch foam lining like the kind used for outdoor camping underneath the bottom sheet. But as with a reflective blanket, moisture buildup can become a problem.

Another way to add warmth to the bed is to add "things" to it to make it as warm as you want it. Some of these things might include hot-water bottles or heated, unopened cans of soup. Hot-water bottles can sometimes be purchased with a furry cloth covering. But a hot-water bottle or metal canteen wrapped in a flannel shirt will do as well. A unique variation of the hot-water bottle, found in some European countries, is the flameproof bag filled with cherry pits. (Could olive pits do as well?) When heated, they act like mini-charcoal briquets, holding and releasing the heat for hours.

Some interesting 19th century hot-water bottles are illustrated in David de Haan's colorful book, *Antique Household Gadgets and Appliances.* One of the most popular was the stoneware hot-water bottle. A ceramic stopper was placed in the center of the loaf-shaped bottle which was always kept stopper-side-up.

"This was because of the difficulty then, and for that matter even now, of making a ceramic screw fit well enough into a ceramic thread to remain watertight. A rubber washer made all the difference, enough to allow a turn-of-the-century porcelain model like the 'Mecca' to be used lying on its side but still with some risk. Even in the 'Marion', an early rubber hot-water bottle, the convention of keeping the stopper upright to avoid leakage was continued. Looking very like a leather brief-case, the bottle was intended to stand by itself. Some manufacturers were being bolder and making the familiar rubber 'lying down' type as early as 1890s, but those customers who had experienced wet beds from leaking bottles were joined by the sceptics and steered well clear of the unreliable rubber hot-water bottle, preferring nice safe metal ones. Made of copper they were mostly round and flat in more or less the same shape as the faithful old warming-pan. For those who preferred to warm their bodies rather than the bedclothes, there was the anatomically-shaped Belly Warmer resembling an oversize hip flask" [1].

The bedwarmer described as the "faithful old warming-pan" was a round metal container attached to the end of a long wooden handle. In 19th-century homes, the container held lumps of hot coals. When passed swiftly back and forth between the sheets, it ironed warmth into them. When not in use, the bedwarmer was hung by its handle beside the fireplace or woodburning stove.

Another way to warm the bed was by using blocks of soft talc stone called soapstone. A piece of soapstone was given to each member of the family and these were heated on the woodburning stove before bedtime. When wrapped in newspaper and then in cloth and carried to bed, they would radiate their stored heat for many hours. (Some wood-burning stoves are faced with soapstone to create this long-lasting heat source.)

One problem with heated stones is that for simple bedwarming they must be hot, but not too hot. One modern backpacker, experimenting with a hot rock carried directly from the campfire to bed found out why. Within a few minutes, the rock had melted through the synthetic material of his sleeping bag, through the floor of the tent and lay smouldering on the ground.

An alternative to heated stones is heated, unopened cans of soup or beans or whatever else is handy on the shelf. When heated (but not overheated or else they will explode) they can then be rolled to the bottom of the bed. While a hot-water bottle, heated stones or small electric heating pad will keep the feet warm for hours, warmed cans of soup will last only twenty minutes or so. But in a pinch, those few minutes may be all you need to coax you into a cold bed.

Hats, gloves and bedsocks will also help. As mentioned earlier, nightcaps were once worn by every member of the family to keep body heat from passing freely out of the head. In some countries, like England, nightcaps and mittens can still be purchased—in matching sets—especially for wearing to bed on cold winter nights.

Another type of bedwarmer, so obvious it is often overlooked, is the human body. "One person in a bed can make a comfortable 'nest.' Two can make it quite cozy. Three can weather any storm. Four can make it unbearably warm. . . ." This custom, practiced in many countries during the winter months, is called "bundling."

Bundling, which flourished for many years in rural New England settlements during Colonial times, was both denied and accepted. For some people living in the northeast, bundling was a simple way of keeping warm: putting "multiples" into a bed raised the heat to a tolerable level. In those days, the traveling salesman was sometimes referred to as the "professional bundler." He provided some verification of the custom, as the tales he brought back tended to confirm suspicions.

For other rural families, bundling was simply a convenient form of courtship. When a young woman of marriageable age was visited by a likely suitor, it made sense to the family to encourage them into bed where they might carry on conversations under the protective comfort of the quilt. This was better, they reasoned, than to insist that the couple remain sitting on the woodbox or in chairs before the fire, after the rest of the members of the household had gone to bed for the night, some perhaps even in the same room.

The custom of bundling was not always known to both parties. Sometimes, before such an event took place, it took a little coaxing or understanding that such a custom was indeed accepted by other members of the community. Usually the offer was made to the man who came to the house for a visit. Sometimes the visitor was a woman.

One such woman described the progress of a romance with her future husband. After several times walking hand in hand with her, he proposed that they "bundle tonight."

"Bundle what?' I asked. 'We will bundle together, said he; you surely know what I mean.' 'I know that our farmers bundle wheat, corn-stalks and hay; do you mean that you want me to help you bundle any of these?' enquired I. 'I mean that I want you to stay with me tonight! It is the custom in this place, when a man stays with a girl— if it is warm weather, for them to throw themselves on the bed, outside of the bed-clothes; if the weather is cold, they crawl under the clothes, then if they have anything to say, they say it—when they get tired of talking, they can go to sleep? This is what we call bundling— now what do you call it in your part of the world?' 'We have no such words,' answered I; 'not amongst respectable people, nor do I think that any people would that either thought themselves respectable, or wished to be thought so.'

"I have since made enquiries about bundling, and have learned that it is *really* the custom here, and that they think no more harm of it, than we do our way of a young couple's sitting up together. I have known an instance, since I have been here, of a girl's taking her sweetheart to a neighbor's house, and asking for a bed for two to lodge in, or rather to *bundle* in. They happened to have company at her father's, so that their beds were all occupied; she thought no harm of it. She and her family are respectable.

"Grandmother says that bundling was a very common thing in our part of the country in old times; that most of the first settlers lived in log-houses; that these log-houses seldom had more than one room that had a fireplace—in this room the old people slept—so that if one of their girls had a sweetheart in the winter, she must either sit with him in the room where her father and mother slept, or take him into her sleeping room—she would choose the latter for the sake of being alone with him; but sometimes when the

cold was very severe, rather than freeze to death, they would crawl under the bed-clothes; and this, after a while, became a habit, a custom, or a fashion"[2].

The custom of bundling was sometimes extended beyond family and potential family members to include neighbors. A fellow who grew up in the Berkshires in Massachusetts during the 1930s recalled that when it got bitterly cold, "We had a standing agreement to get together with the neighbors.

"There were about twenty children and we'd shove the beds together and crawl in. There weren't too many of us boys, so when I crawled into bed it was like sleeping in a girls' dormitory. I think my parents enjoyed their arrangement too.

When I was in college, I once went home with a friend on a winter break. They didn't heat the bedrooms, so we slept in the same bed. No one thought a thing about it. Nowdays, of course, people would look at it differently: two fellows sleeping in the same bed. . ."[3].

REFERENCES AND NOTES

1. David de Haan, *Antique Household Gadgets and Appliances,* (p. 147) Blandford Press, U.K./Barron's, U.S.A.
2. Nathan C. Fuller, *The Down East Reader,* Selections from the Magazine of Maine (Philadelphia, Lippincott, 1962)
3. Mad River Jack, from a private conversation with a backwoods Vermonter, July 1979.

THE

INTERIOR

"In time . . . in an effort to simplify angles and lines, the human frame was often forgotten. Human comfort was forgotten too as the upholstered chair was replaced by furniture made from materials that, reptilian-like, assumed the temperature of their surroundings."

Furniture, according to Webster, is an article of convenience or decoration. That familiar table, lamp, chair or rug greets our eye as we enter a room, welcoming us like an old friend or maintaining a "look but don't touch" attitude.

Made of wood, wicker, fiber, bamboo, textiles, metals, synthetics, glass . . . furniture can also leave us feeling hot or cold. It can passively or actively maintain, add to or subtract heat from a body. A woven-wire chair such as those often found in an ice cream parlor will elevate the body and expose it to cooling drafts. An overstuffed chair will surround the body like the comfortable lap of a buxomy aunt, holding warmth and giving it back.

Some furniture can also give the *appearance* of being warm or cold. Brown, orange and maroon tones will convey a psychological sense of warmth. White, green and blue will not. Similarly, a cluttered room will seem warmer than one containing only a few items of furniture.

In the 19th century, the cluttered Victorian parlor had a sense of warmth. It was filled with upholstered chairs and footstools. It contained tables of various sizes, candleholders and candelabra, gas or electric lamps, floral-designed rugs and floral-papered walls, hanging curtains and the presence of people from portraits hung on the walls. One sensed a crowd in the room, sensed in the clutter, the warmth of them.

In medieval times, that sense of warmth was present simply because all the furniture—unlike some of our contemporary molded fiberglass, plastic or aluminum frames—was carved from warm-toned woods or made from tooled leather or natural fibers.

Tapestries—like the furs hung on the walls of prehistoric caves—were used to provide additional warmth to the rooms. In time, as the simple woven textiles evolved into great pictorial works or art, their functional use was overlooked and they became primarily objects of decoration.

Throughout the centuries, a heavily woven cloth was hung over the entryway to some rooms. Called portieres in the 1800s, these floor-length curtains blocked the passage of cold air into and out of the room. They could also be attached to the back of a door frame by a "portiere rod" which would swing up and to the side whenever the door was open. The heavy curtain sealed out small cold air leaks when the door was closed.

As with curtains that covered the windows, portieres that draped the doorways often had a valance box made of matching materials or of carved wood fastened above them.

While the valance box may have served a strictly decorative function—to hide the top edges of the curtain and curtain rod—it also stopped the convection of cold air into the room.

Where the walls were not covered with tapestries or curtains, they were frequently panelled with wood or painted with elaborate murals. Frescos, made with colored paint worked into wet plaster, also decorated the walls.

In Dutch homes, in particular, the wooden doors, walls, cupboards, window frames and wooden shutters were often painted with colorful designs. In time, through the introduction of wallpaper in the early 16th century and its later mass production for popular use, the walls and ceilings of even the most modest house could be covered with colored designs or pictures that evoked a sense of warmth.

In early New England, families rarely had more than two chairs per household. Seating was supplied by long wooden benches, storage chests or small wooden stools.

Some wooden benches called settles—the forerunners of today today's sofa or couch—were heavy enough to be nearly immovable. The settle had a high back and an overhead hood which protected those who sat upon it from winter drafts.

These semi-enclosed benches, sometimes divided further into separate seat compartments or into "topless" boxes, were built into some of the churches, such as North Church in Boston, to help block the drafts of cold air and to keep the parishioners warm.

One special kind of settle that has, as an antique, found a number of new uses, was called the Bacon Cupboard. In self-sufficient rural homes, the pioneer wife often salted and smoked sides of bacon and ham in the large central fireplace. The cured meats were then hung on hooks inside a large wooden cupboard—the Bacon Cupboard—which had been placed next to the fire. The cupboard had a storage chest extending from its base. The hinged top or set of drawers could serve as a seat before the fire with the cupboard acting as a backrest.

Like the Bacon Cupboard, another settle, the hutch table, also had multiple uses. The table, designed for both eating and sitting, contained a hidden chair beheath the top. When the top was lifted, it provided a high, draft-resistant back for the chair that was formed by the four legs and center board. When placed before the fire, the hutch, like the Bacon Cupboard, formed a formidable barrier for the heat radiating from the fireplace and kept the person sitting in it quite warm.

To regulate the heat from the fireplace, some people found it necessary to use a firescreen, a double-glass shield, mounted on an adjustable rod. The glass shield enclosed two framed tapestries which when extended away from the glass to either side directed the heat away from the person sitting in the chair, and exposed a view of the fire through the double-paned glass.

In the Middle Ages, the single chair with a back and side arms was reserved for the master of the house. (It was from this custom of reserving the chair for the master or for a distinguished guest that the term "chairman" evolved.)

By the mid-1600s these chairs began to appear with horsehair or wool pads tied to their backs and seats. These upholstered chairs added a previously unknown "cushioned comfort" to the functional chair as well as a new element of warmth.

In the 19th century, the era of the easy chair began when the coil spring, the principle factor in modern upholstery, was adopted. At first the easy chair was used by the elderly family members or the infirm and then, typically, only in bedrooms. It allowed them to sit in comfort away from their beds.

Because of this special function, most easy chairs

were accompanied by a small footstool or a frame which could contain a chamber pot. The chair, like other wingbacked styles, when drawn close to the fireplace would reflect the fire's heat back to the person sitting in it.

Its high back protected the person from cold drafts while the footstool lifted his or her feet off the cold floor.

Another way of keeping the feet warm was by using a footwarmer,

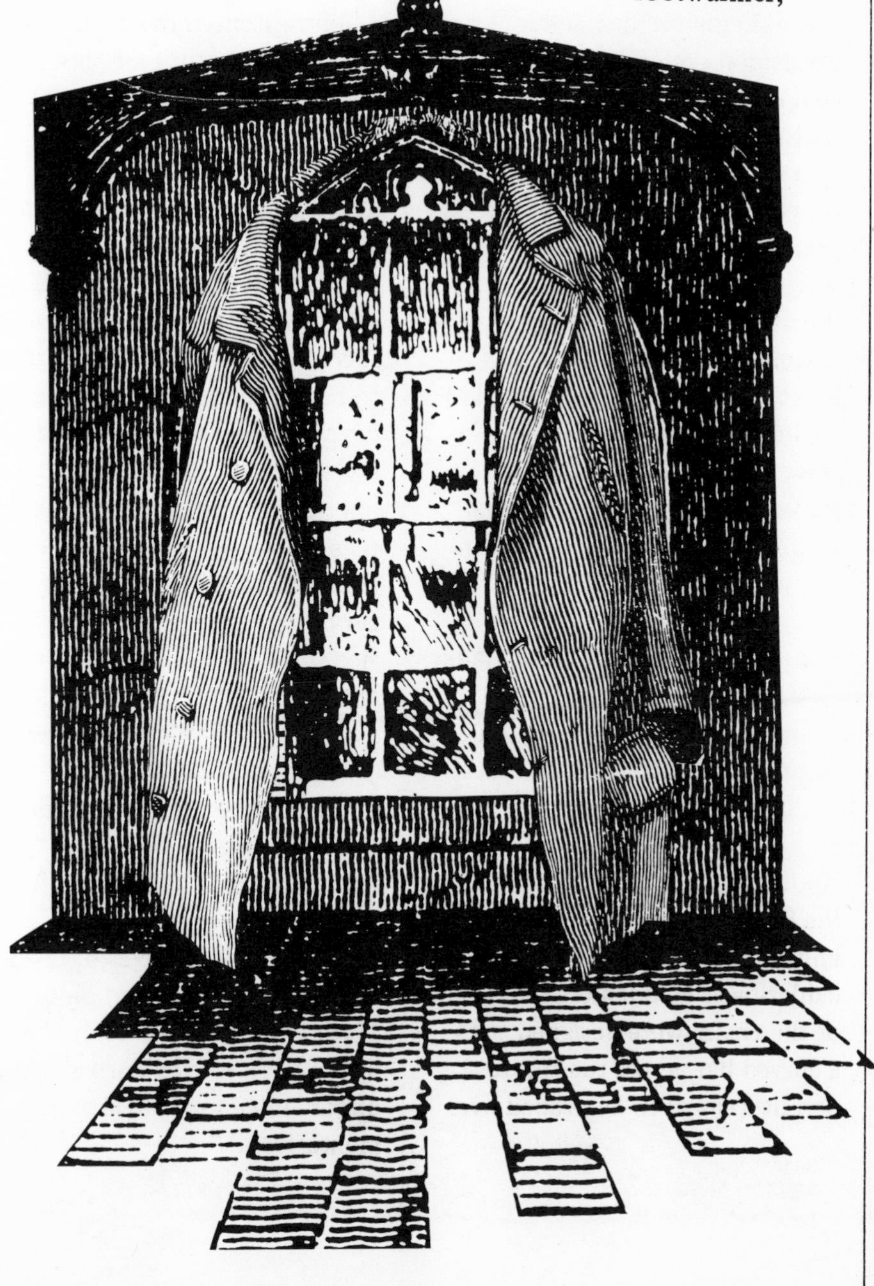

a small metallic box designed to carry lumps of hot coal from the fire to wherever it was needed. One footwarmer, advertised in the Sears, Roebuck & Company catalog in the mid-1800s, was called the perfect footwarmer. Made of heavy galvanized metal, it was 20 inches long and weighed 7¼ pounds. A perforated metallic drawer held the hot coals. The footwarmer was covered with an "extra quality" velvet Brussels carpet and sold for $2.95.

A cold-weather item that sometimes accompanied the upholstered chair was a 12-inch by 12-inch framed cloth on a swivel. It was attached to the top of the chair back and could be swung around to protect the head from any unexpected drafts.

The skirts on furniture legs, like the floor-length folds of a woman's skirt, reduced the movement of cold air around the base of chairs and couches.

Another unique draft shield was the Shakers' crocheted or knitted "sleeves" for their ladder-back wooden chairs. Some of the chairs were designed with a curved bar across the top to permit a shawl to be draped over the back. The shawl served much the same thermal function as the sleeve.

Although the Shakers continued to use straight-backed functional wooden chairs, the comfortable upholstered chair and the sofa, a luxury item in the 18th and early 19th centuries, at last became commonplace in almost every household.

The transition from easy chair to overstuffed furniture was a simple one. Then, ironically, after reaching a peak of comfort, these comfortable chairs began to be stripped of their spring coils and padding. Furniture styles began to reflect the introduction of new synthetic materials on the market. Just as clothing styles of the mid-20th century emphasized the new synthetic fabrics which offered conveniences (and problems) that the natural materials did not, so too did contemporary furniture styles. The new styles included the use of metals such as aluminum and tubular steel, plastics, rubber and synthetics in place of woods such as oak, black walnut, mahogany and pine. These new materials offered a lightweight convenience and the opportunity for radical experimentation.

Chairs at last could offer a place to sit as well as a sculpted form. Molded of fiberglass, clear plastic and aluminum, framed out of sheet metal and vinyl-coated steel wire, furniture as an art form began to replace the easy chair. In time, as the new materials were pushed to their limits, and in an effort to simplify angles and lines, the human frame was often forgotten.

Human comfort was forgotten too as furniture was made from materials that, reptilian-like, assume the temperature of their surroundings. In a room where everything is warmed to the same degree, they feel warm. In a home where the temperature is lowered or the central heating system turned off, they feel icy to the touch.

It doesn't take long—living through a winter with little or no heat—to discover that this feeling can be reversed. Some furniture can be covered with warm-toned materials to give the appearance of warmth.

A glass-top table—or any table for that matter—when covered with a cloth in the color range of orange, brown and maroon will seem warm. Lit at dinnertime by flickering candlelight, the atmosphere around the table will become a haven from cold winter weather.

Similarly, a brown-colored vinyl lounge chair with the "look of real leather" that looks warm but feels cool, can be covered with a brown slipcover that has a thick, loose weave or a heavy nap, like velvet, to make it look and feel warmer. As with clothing, the thickness of the fabric will trap body heat and keep the person sitting in it warmer than will a chair or sofa with a tightly woven or smooth surface.

In addition to slipcovers on sofas and chairs and warm-toned coverings for tables, other measures can be taken to increase the look or feeling of warmth in a favorite room.

A cluster of pillows or pillow furniture will surround the body and keep it warm as will an overstuffed chair. The furniture can be drawn together in one part of a room to create—clutter fashion—a circle of warmth. Like a crowd of people, a cluster of furniture gives a sense of warmth.

In addition to increasing the warmth of traditional or contemporary furniture, developing new furniture designs can stretch the imagination and open up unique people-heating possibilities. One innovative design, for example, is a sofa frame upon which have been sewn hundreds of sea anemone-like tentacles. Though this design may never find its way into mass production, one can imagine being swallowed up by its tentacled warmth. However, it could be a problem trying to carry on a conversation with someone else in the room or finding a firm surface on which to push up and out of the couch.

Another imaginative furniture design is the woven hanging on display as an art object in the textile division of the Smithsonian Museum of History and Technology. Entitled "Meditation Space", the hanging consists of three hammock-like "cells" [1] suspended like a cluster of grapes from the ceiling To reach them, one enters from an oval opening at the front. Once inside, one can slip into a comfortable niche for sitting by moving to the cell at the left, right or back. The hanging is somewhat flame-shaped when viewed from the outside, and it is easy to assume its hidden warmth: warmth provided not only from the use of natural fibers and from the enclosed space, but also from the people sitting inside.

Temporary or permanent wall coverings can also provide a greater sense of comfort. Walls can be wood-panelled, wallpapered or painted with warm-toned colors. A photomural of a warm weather scene can be used to cover one wall, or the wall can be lined with bookshelves. Paperback, cloth or leather-bound books absorb heat and act as a uniquely pleasant insulation.

Colorfully painted folding screens or moveable room dividers can enclose an area and add warm tones. Woven wall hangings and macrame, now sold as art objects, can, like the tapestries of medieval times, serve as "clothing" for the walls. Oriental rugs or thickly woven, floral or warm-toned rugs can be laid—in layers if desired—on the floor. An old-fashioned, braided wool rug can be laid on top of a room-sized or oriental rug. These can then be moved into storage, if desired, in the spring.

In older buildings, where the ceilings are 14 feet or higher from the floor, warm air drifts upward out of reach. A small rotating fan can be used to recirculate the air downward, or the ceiling can be lowered to force the warm air down into a smaller space. Heat can also be drawn downward by a heat syphon or recycler [2]. With wallpaper or paint, the ceiling can even be made to seem lower than it actually is. Gray or blue paint used on the ceiling and extended a few feet down the walls will give the impression that the ceiling is lower. This illusion will make the room seem warmer.

In addition, clear glass lightbulbs can be replaced in some general lighting fixtures with cream, orange or coral-colored bulbs. Again, this provides the appearance of warmth without actually creating it. To create a real sense of warmth from the lighting without adding more electric lights to the room, an Aladdin lamp or candles placed in candleholders in front of mirrors can be used. The use of oil or kerosene lamps or candlelight will add heat from the flame as well as an old-fashioned coziness.

Anyone who has ever read by the light of an oil lamp knows how much heat is given off at the top of the glass chimney. On a wintery night, an oil lamp can be an unexpected source of warmth. With a simple trivet attached to the base of the lamp and centered above the glass chimney, the lamp can serve two purposes.

In the mid 1800s paraffin (considered an auxiliary fuel to wood or coal) and later oil lamps were often used with trivets. The trivet could hold a small pot of tea or heat a small amount of food for the person sitting beside it. This simple dual-purpose, heat-generating lamp became the forerunner of the paraffin stove which was a popular household item for over 60 years, until the 1930s.

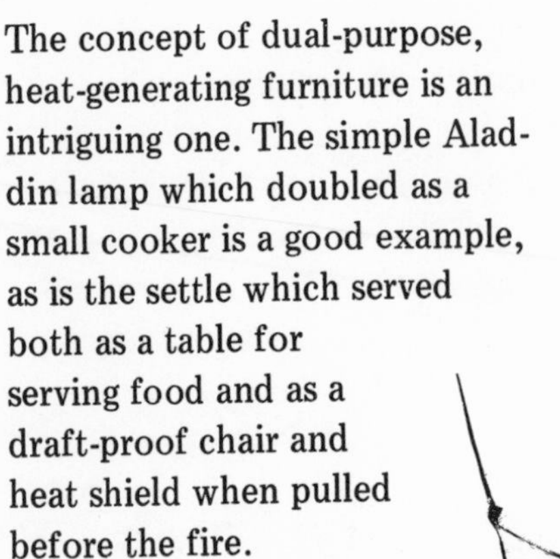

The concept of dual-purpose, heat-generating furniture is an intriguing one. The simple Aladdin lamp which doubled as a small cooker is a good example, as is the settle which served both as a table for serving food and as a draft-proof chair and heat shield when pulled before the fire.

Another dual-purpose type of furniture that may become commonplace in homes of the future, and which takes advantage of small-scale solar and renewable resource technologies is the solar couch [3]. The solar couch is the conceptual brainchild of Dr. William Shurcliff, author of many books on solar energy. Although the couch has been presented only in design stage at the time of this writing,

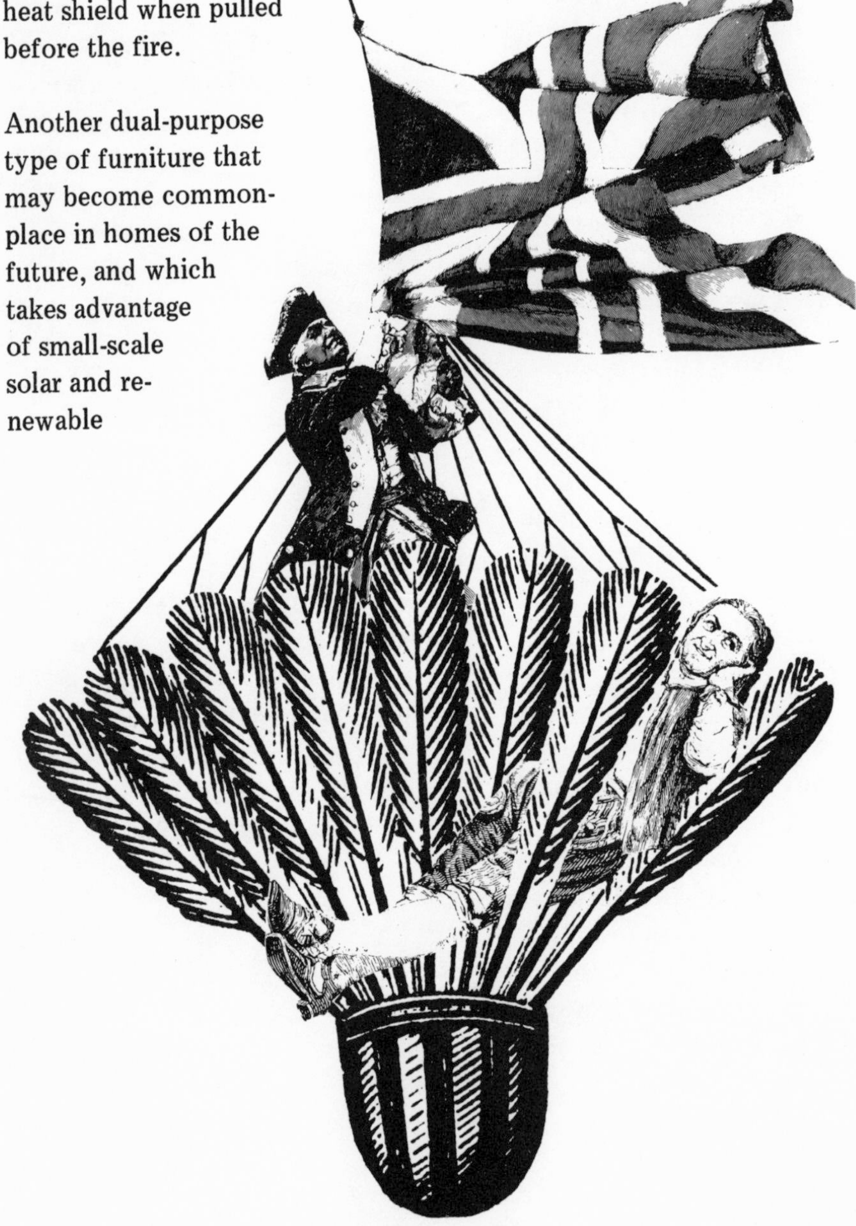

a description of the proposed scheme may stimulate some readers to design solar couches of their own.

Shurcliff's solar couch consists of a water storage tank or series of barrels placed in front of a south-facing window. The storage unit is covered by an insulated cushion and serves as a window seat or couch. On a day when the winter sun shines directly in the window, the seat of the couch is raised and the storage barrels beneath it are exposed to the sunlight. The cushioned back of the seat also has an attached insulated reflective shield which focuses additional sunlight down upon the storage system. Sunlight heats the water and when the cushion is reset, the sofa becomes both a localized heat source—sitting on it is like sitting on a warm radiator—and part of the hot water heating system.

CENTRAL

KEEPING

"I'd like to live in a house that was so well insulated that all it would take to heat it would be the daily burning of *The New York Times*." William Shurcliff

Clothing provides one method of localizing heat. Another method of focusing heat is to set boundaries beyond which the heat will not or cannot travel. A single room in the house can be selected for this purpose.

Most houses have a heart: a single room that is either centrally located or, being the most comfortable, is most often used. In Colonial times, the kitchen or kitchen "keeping" room provided a place where the majority of family activities took place. In most cases this was because the central commons area *was* the house.

In time, to deal with the winter cold, some easterners discovered that they could avoid the out-of-doors, get from here to there, from the house to the tool shed, for example, by attaching the shed snug against one wall of the house.

If they had a field which gave them crops to store for the winter, they might add another room, a storeroom, beyond the first. And to save the trouble of wading through deep snow to tend to farm animals or to run to the privy, they might add a barn and an indoor toilet further down the line.

Barns attached to houses were nothing new. In Europe, many farmers herded their animals into a winter shelter beneath the house and tended to them there. Milk, cheese, eggs, butter and, of course, heat from the warm-blooded animals could be drawn upstairs into the main house.

In time, wherever the rooms of Colonial America grew, whether they were built to reflect expanding farm capacity or to accommodate the expanding family, technology soon followed.

In the 19th century, gas, and later, electric lights lit up the most distant corners of the house. In addition, pipes for carrying water ran between the rooms. To keep the pipes from freezing, continuous heat was needed during the winter months.

When heat from multiple woodstoves or the fireplace (which provided heat for a room and sucked it out again) proved unsatisfactory, the central heating system was invented.

Central heating! With its insidious metallic arms linking every room in the house but uncoupling the occupants at the same time. A well-intentioned technology but one that brought about the "dismemberment" of the family unit.

With the advent of steady, dry air in every room, people could move away from one another. They could become antisocial. They could be warmed from both the front and back at the same time. A closed door, therefore, could stay closed as long as the person inside had charge of the air vent and could regulate the heating or cooling of the room.

Central heating marked a technological advance but the social loss was enormous. Gone were the kitchen helpers and constant companions, the nightly family musical sessions, religious or philosophic studies, story telling and recitations. The family unit in winter had disappeared, members retreating to their separately heated caves. They needn't return to the kitchen until it was time to eat again.

Along with the disappearing family came disappearing walls. By the 1840s technology had found a way to simplify or "balloon" the wall structure.

Walls no longer served as load-bearing monuments to finely crafted, heat-absorbing masonry, or bore the hard-earned yet prideful scars of roughly hand-hewn, insulating logs. They be-

came drafty skeletons of earlier wall structures, air shafts in which air was actually drawn up into the attic and out of the house.

Interior walls became skeletons of wood with thin paper skins nailed to them. Running like veins inside these walls was a system of pipes from the furnace. The pipes carried invisible heat from the heart of the house into every room: unseen, seemingly indispensable, . . .yet vulnerable.

For, in time, the system became dependent not on resources provided firsthand and stored at the house, but on domestic and foreign fuel sources stockpiled and converted to usable energy someplace else.

Although our contemporary houses have now spread in all directions: looking aerially like giant *S*-, *T*-, or *L*-shaped structures, it makes sense to reconsider our roots. To conserve expensive fuels, we might select a single room to serve, once again, as the central keeping room. This room could be designed to serve multiple functions and the rest of the house could be kept cooler or closed off during the winter months.

As illustrated ingeniously in *Living in One Room*, a book directed primarily at apartment dwellers, this central room could have all the warmth and comfort of "home" [1]. At the same time, the rest of the house could be made comfortable by relying on limited or focused heat, quilts or extra layers of clothing.

The problem with shutting down rooms in houses is that many have some insulation in their outer walls but none in the interior walls. Because of this, localizing heat in a single room is somewhat complicated. If the heat cannot be trapped or bounded sufficiently within the room, it seeps through the walls, doors, ceiling and floors. Insulation, therefore, is the first step in centralizing heat.

Insulating a room can provide a temporary or a permanent "fix." The materials can be manufactured and purchased from a store or they may be items close at hand.

In earlier times, walls made of sturdy masonry, adobe, hand-hewn timber or half timber and half mud and woven stick (wattle-and-daub) provided insulation naturally. Where additional insulation was needed, it was made from materials that were close at hand. These natural insulation materials included sawdust, seaweed, newspaper, rags, moss, birch bark, woven mud and twigs, pieces of shale, layers of corn husks or straw, thickly woven eel grass and layers of cleaned cattle hair placed between two sheets of paper-like material.

In early Pennsylvania settlements, the cracks in the log frame houses—there were 10 hardwood logs to one house—were stopped with "slender pieces of chinkin' split out of basswood."

"We mixed up some clay mud and plastered over the cracks and chinkin' on the outside. The inside of the walls we'd hew down smooth enough so we could paste paper onto 'em. The floor was of split basswood smoothed off with an adz. Our fireplace was of stone, and was so large we could roll logs eight feet long right into it. Those logs were big and green enough to last for days. We had no candles, no lamps, no nothing, except the light from the fire in the fireplace" [2].

In those days, a child with a small pitcher of water could pour a few drops of liquid into the cracks in the window frame and create an ice seal which would keep out the drafts. Sometimes the ice sealed more than the windows. For one hardy pioneer living the winter alone in her small cabin, the ice nearly sealed her doom.

"Last week . . . I got sorta scared here," she explained to a visitor who arrived after the ordeal was over. "I 'bout give out a bit of anything t'eat. See, all th' meat 'n stuff's in th' cellar except th' dried beans. Well, you can't eat dried beans in a minute. Y' have t' have time t' cook 'em. Well I couldn't get th' cellar door open. It come a freezin' spell an' that door swelled up. Well I tried ever'day t' open the door, an' I got a rock an' I tried t' rock it down 'til I could open it. Even had Irish 'taters in there and couldn't get Irish 'tater t' eat. I's sorta scared I tell you. You can't get nary hand on nary thing t' eat, cook. I can cook but I couldn't get m' hands on 'to cook. So I said, "Well, I reckon next thing best t' do's t' boil a big kettle a'water. . . .

"First thing I done when I went in th'kitchen this mornin's t' go see if my door'd open so I have something t' cook fer dinner, an' it opened just as pretty as it could be. Lord, I rejoiced! I did. I rejoiced an' run in there an' got me a can a' beans an' put on t' cook" [3].

Today, using modern insulating materials and paying particular attention to the small leaks and cracks, we can keep the cold weather out. However, while these materials are far better than those used in the past, poor application can eliminate most of their benefits. Careful attention to small air leaks, perhaps most important, is the most time-consuming and thus the most overlooked task in sealing a house. A dozen small air leaks can add up to one large draft.

One way to locate air leaks in a house is to hold a candle near the suspected source. The flame should tell you whether there is a draft. Another way is to attach lightweight strips of plastic to the curved part of a coat hanger. The rest of the hanger can be shaped into a long handle. This *draft detector*, simple enough for a child to make and use, is especially useful in locating air leaks in hard to reach places.

Walls, ceilings and floors can be insulated to achieve a certain "R-value", or thermal resistance value. Materials can be purchased to caulk or weatherstrip around the inside of door and window frames to seal out the cold air.

The fireplace can be temporarily plugged with newspapers to seal any air leaks around the damper. Styrofoam and plastic dry cleaner bags can also be used, or a "cap" which is manually raised or lowered from the outside can be obtained for the top of the chimney. The problem with styrofoam and plastic is that if a fire is accidentally lit before the damper is opened, as sometimes happens, the styrofoam or plastic will release deadly fumes into the room.

The fireplace can be permanently sealed by closing it off with sheet metal or bricks. (Some companies now offer a combination woodburning stove and shield, designed to seal off and convert the fireplace with a one-step installation.)

When a woodburning stove is placed in front of the fireplace its fluepipe can extend up the chimney. When bricks are used as a seal they will serve as a heat sink to store the heat generated by the stove and slowly release it. A metal shield will reflect the heat from the stove back into the room, but will not store it. A sheet of metal or aluminum foil placed behind steam-heat radiators will create the same increased radiation effect.

Because over 50 percent of the heat in a room can pass through a single-paned window and through the cracks around its edge at night, window insulation and caulking is essential. Storm windows can be purchased ready-made, made to order, or they can be a do-it-yourself project.

Window shades are effective when used in combination with storm windows. Simple window shades, in fact, are more effective than Venetian blinds and draperies in sealing out the cold air. An exception is the foam-backed Venetian blind sealed tightly against the window frame.

Unless the curtains are heavily draped at the edges and closed off by a valance box at the top, they may actually increase the circulation of cold air in the room. The curtain creates a fabric "conduit" around which the cold air can circulate. As the warmer room air picks up the cooler air from the window pane, it drops toward the floor, drawing the warmer air at the top of the room down behind the curtain. A valance box or a strip of fabric tacked to the window frame and draped over the top of the curtain will stop this cold air cycle.

Insulated window shades such as the IS High "R" shade made by the Insulating Shade Company of Guilford, Connecticut, are better than curtains or Venetian blinds. These shades are made of multiple layers of aluminized Mylar and dividers attached to a roller shade. The Mylar sheets separate when the shade is pulled down, creating a series of mini-air pockets. Because the effectiveness of an insulated window shade depends on how tightly it fits against a window frame, an enclosed track is attached to each window frame to hold the shade in place and prevent air from escaping around the edges or bottom. The trapped still air becomes the insulation.

An insulated wooden shutter set into the window jam, closed at night and swung open in the daytime is another option. In a pinch, heat loss through the windows can be brought down to near zero at little cost, by using styrofoam or sheets of Thermax, a Celotex insulating product. Thermax, with heat-reflecting aluminum on one side (which can be painted to match room colors) can be cut with a sharp knife or carpenter's saw to fit each window. Depending on size, the cost can run between $3 to $5 for an average window. Clear plastic can be used as well.

If sheets of foam are used, they can be placed against the window at night and secured by tacks or with special "Nightwall Clips" [4]. There are two drawbacks to using foam plugs for windows: they must be stored in the daytime (unless the seal is a "permanent" one for an obscure window), and in the event of a fire, as mentioned earlier, the foam would release toxic fumes into a room.

Doors must also be weather-stripped and well-sealed. In some cold areas of the country, temporary walls are erected in the wintertime to create an unheated vestibule outside the main entry door. This keeps the cold air from replacing the warmer air inside the house each time the door is opened. A closed-in porch will serve the same function.

When the door is closed, air leaks at the bottom can be reduced or eliminated by one of several insulation methods. One is the use of a vinyl strip backed by aluminum, called a sweep. The

other is a vinyl bulb threshold which attaches to an aluminum threshold mounted permanently to the floor.

Drafts can also be controlled by placing an *energy eel* against the base of the door. The energy eel is an old-fashioned heat saver. When not in use, it is usually hung on a hook by the door. When the door is closed, the eel—a long fabric tube filled with sand—can be tossed against the base of the door. It assumes whatever shape is needed to seal out the air. A hanging loop is attached to one end of the tube. When eyes are sewn on the other end, it becomes an "eel."

If insulating measures are carried to an extreme, a house can become so well sealed that it would achieve a "thermos" or "vacuum bottle" effect. It could then be warmed by little more than the heat from a candle. (In such a case, providing adequate ventilation might become the problem.)

In the Saskatchewan Conservation House in Regina, all ventilation is provided by a simple air-to-air heat exchanger which recovers 70 percent of the heat from the used air before it is vented from the house. Inside a simply constructed box, a thin film of polyethylene separates the incoming from the outgoing air in the heat exchange unit. In this manner, the incoming air is warmed by the exhausted air before it enters the house.

Hypothetically, a room could be lined with aluminum foil to seal out the cooler air and reflect heat back into it. The heat from a single lamp focused on one wall could then be reflected continuously off all the walls and ceilings, warming a person sitting naked in the center of the room.

This hypothetical room has a realistic counterpart. Some Eskimos, though not living in igloos lined with aluminum foil, live in highly reflective ice-lined rooms. Each snow block contains millions of trapped air cells which make excellent insulation. The heat of a simple oil lamp and the heat from their bodies keeps the Eskimos warm in temperatures well below zero.

Since most of us live in shelters that are not so well sealed, we have to consider the necessity of properly insulating them and at the same time consider centering the heat in a single room.

The draft-free central comfort room can be designed to accommodate the family's needs in winter. The more clearly defined those needs, the easier it is to create a room that is not just liveable, but fun to live in.

Moveable dividers, ceiling-to-floor curtains or screens can temporarily divide a room. Dual-purpose furniture that can be "hidden" or converted to serve multiple functions and that looks and feels warm can be used (see section on furniture). The trick is to make the room both *usable* and comfortable.

Before moving on to a discussion of what sources of heat could be used in this central comfort room, it would be well to remember that like everything else in the world, even a well-insulated house can have its critics.

Nearly 130 years ago, Catherine Stowe was warning readers of *The American Woman's Home* about the dangers of a well-sealed home. "It's a terrible thing to reflect upon," she would tell them, "that our northern Winters last from November to May—six months long—in which every window crack has been carefully caulked to make it airtight, where an airtight stove keeps the atmosphere at a temperature between 80-90 degrees. . . .Better, far better, the old houses of olden time with their great roaring fires and their bedrooms where the snow came in and the wintry winds whistled. Then to be sure, you froze your back while you burned your face —your water froze nightly in your pitcher . . . but you woke full of life and vigor" [5].

Of course, any measure carried to an extreme can create new problems while relieving the old ones. But, if given a preference, most people would prefer to err on the side of a well-sealed house rather than a drafty one.

REFERENCES AND NOTES

1. Jon Naar and Molly Siple, *Living in One Room* (New York: Vintage Books, 1976)
2. Clifton Johnson, *What They Say in New England and Other American Folklore* (New York: Columbia University Press, 1963), p. 275.
3. Eliot Wigginton, *Foxfire* (Garden City, NJ: Doubleday and Co., 1972), p. 26.
4. Zomeworks Corp., P.O. Box 712, Alburquerque, NM 87103.
5. Catherine Beecher and Harriet Beecher Stowe, *The American Woman's Home.* 1860. Reprint. (New York: Arno Press and New York Times, J. B. Ford Company, 1971).

WITHOUT

CENTRAL

"We are too much inclined to believe that because things have long been done in a certain way that that is the best way to do them. . . .At times, the only thing to do is to cut loose and do the unexpected! It takes more even than imagination to be progressive. It takes vision and courage."
Norman Bel Geddes, *Horizons.*

The pattern of fuel scarcity and soaring prices is not new to this country. Fuel shortages in the past have forced transitions in the ways we have heated our houses. We have made the transition from wood heating systems to those using coal, oil and natural gas. We have changed technologies: from the fireplace to the closed woodburning stove to the central heating system. In cyclic fashion we are in the process of transition once again.

In learning how to focus heat and direct it toward those places where it is needed, we may find our central heating system no longer serves a useful function. Is this impossible to believe? Not at all if one considers, for example, the current state of solar technology. In a well-designed solar or intensive, passive solar home, a central furnace is usually installed to satisfy the requirements for a bank loan, not to supplement the heating needs of the homeowner.

Most of us, however, do not live in solar homes, but we can break our dependency on the central heating system. We can browse through this book once again, consider our current living space, then select those methods for keeping warm that promise to give us the most control over our personal comfort while fitting our uniquely individual lifestyle.

Historically, transitions in the methods of keeping warm have been initiated by the wealthier members of society, who can afford to assume the risk of experimentation with the latest technologies. They can afford to innovate, to set the new trends that—where successful—find their way into mass-market production. Instead of doing without, those who can afford to, learn by experimentation how to do things differently. They heat with eutectic tiles or with flue-pipes set into channels beneath the main floor. Instead of settling for an ordinary chair, they settle back into the warmth of a specially-designed solar couch, or sit neckdeep, soaking up the heat in a solarium hot tub. . .a tub which can also serve as a heat storage medium for a solar hot water system.

However, most of the ideas catalogued in this book require little or no money to design warmth back into our lives. They simply require dusting off old methods which worked well in the past and setting our ingenuity to work, and in some cases using modern materials too, to get even better results. Keeping warm by dressing in layers, wearing a cap to bed, clustering furniture away from a cold wall or window, hanging a colorful wall hanging or quilt behind a favorite chair, adding the glowing heat from an Aladdin lamp are all measures of little or no extra cost.

As the book well illustrates, we can keep warm in winter without depending upon complex centralized technologies. Applied commonsense will suffice. And if, in designing ways to keep ourselves warm, we find ourselves devising ways to uncouple our houses structurally from the central heating system, cut back extended plumbing, insulate interior walls and retrofit the central core for comfort and warmth, it is good to remember that such decisions are not backward steps. For the forward progress of humanity does not always have to follow milestones placed one ahead of another in a single direction. Progress can be made as well by turning around and stepping forward in a new direction: retracing steps already made, building pyramid fashion upon the ideas of the past.

If we choose, we can begin now.

Man is addicted to the present. How easy it is for us to do the expected, the predictable, to do as the neighbors do, as others expect us to do.

Man is also addicted to the belief that the present is unlike the past; that the past is obsolete and the future can't be dealt with because it is not yet here. But, in truth, both the past and the future are part of the present. They are as continuously connected to the present as a railroad track is connected to the horizon: always stretching before and behind us out of sight. The present—our awareness of the present—is that mental "train" that we place on the tracks at will.

That we are at any moment leaving the past and travelling into the future can be illustrated by the simple actions we engage in to warm ourselves when the environment is cooler than our liking. Unlike coldblooded animals that adapt to the fluctuating temperatures of their surroundings, thin-skinned homo sapien can take only a slight change in body temperature (a degree or two) before he is in a state of discomfort. Thus when we notice that we feel cool on the skin surface, we experience that sensation in the present.

We use our imaginations to recall things that made us warm in the past and imagine again the warmth that we will soon be feeling, mentally travelling backward and forward in time. But when we take that future action—putting on a sweater, for example—the initial experience of being cold becomes part of the past.

Therefore, because we are always preoccupied with keeping a comfortable temperature about us, we do ourselves a disservice if we allow ourselves to believe that our present actions at keeping warm are in any way removed from the past. And we lock ourselves into a fateful rigidity against change if we pretend not to notice that the future is always with us, passing into the present even as these words are being read.

But how does all this relate to keeping ourselves warm in wintertime? Or to the challenge implicit in the final chapter: living without a central heating system? The connection comes simply from our ability to be aware of the constant forward motion from the past into the future. And from our awareness that those who scoff at historic methods for keeping warm, preferring the "security" of living in the present, have missed the point: even our present methods (central heating, for example) are subject to change, and move into an immediate past that we can draw upon while we continue our movement forward.

Those who cling to present methods for keeping warm do the expected, the predictable. And when these methods include central heating, they leave themselves vulnerable to inevitable higher prices and fear of fuel cutoffs. Those who do the unexpected, are not necessarily embodied with rare courage, rather they are people who have become aware that events are constantly changing, that the present methods of keeping warm are simply extensions and refinements of methods practiced in the past. . .that being human, and thus prone to error, we sometimes branch off onto a side track, a detour, and, recognizing this, must place ourselves back on the main line.

Central heating, it can be debated, is like a side track, or detour. The belief that all rooms must be heated uniformly to warm the individuals in the house is supported by a technology that depends on an infinite supply of fuel. Where that fuel is fossil fuel, an eventual break in our reliance on it is inevitable.

However, living without central heat is not new. People have been living without it for thousands of years. Many people in the world today live without such a technology. Third world citizens come immediately to mind. But even in the recent past—less than 20 years ago—in industrialized England, only about 10 percent of the owner-occupied houses had central heat. Today in America, those who live without a central heating furnace by design, live in solar-heated homes.

Still, for most Americans the central furnace, with our dependence on it, has assumed the symbolic authority of the ancient "sacred fire." Although we consider ourselves long removed from our savage ancestors, our dependence on the system demands that it always be "lit." Its loss is treated as an evil omen; the technical gods must have something terrible in store for us. In our dependent state, we pay financial homage to keep it fueled. We pay psychological homage to its most visible part, the thermostat, and by its degrees, determine our personal comfort.

We have allowed ourselves to become trapped in a situation which gives us the illusion of freedom while the opposite is true. We have allowed ourselves to become what sociologists like to call "neoprimitive man trapped in a technical environment."

Changing our patterns and habits, therefore, to become more self-reliant requires transition steps. But if we remember that we are always in transition, that awareness is one essential element to facilitate change.

In the chapters that precede this section, emphasis has been placed on day-to-day changes we can make in our lives. What follows here is a brief overview of some historic examples of domestic heating and shelter designs. To avoid the pitfall of assuming that because they are historic, these designs are locked into the past, I have taken the liberty of drawing some parallels to contemporary habits in America and to some future visions. But in reviewing the following pages, don't allow yourself to be limited by the author's imagination. Throughout the book, the trick is to lift the *essence* of an idea and adapt it to your own lifestyle.

And, once again, even when adapting an idea, allow it the flexibility to respond to change. If we assume a sense of rigidity (for our own security?) the idea will begin to dictate that all events relate to it and, rather than accommodating our needs, will dictate them. So there must always be a balance point between permanency and flexibility. This final section will only hint at that balance, not provide the answer.

CAVE TECHNOLOGY

There was a time when the highest state of the art for home building was simply finding a suitable cave for a dwelling and then rearranging the stones or carving out niches in the interior walls to provide space for essential household possessions or for sleeping.

While man has always turned to the earth as a source of fertility, prosperity and survival, the cave or earth dwelling for primitive man became a projection of the unity between himself and the earth. As the earth controlled his past, present and future, so too did his close connection with it bring about a centering between man and nature, between the building and its natural setting.

Variations on the cave dwelling theme have gone uninterrupted since prehistoric times. Every culture has its counterpart of the ancient cave. In frontier America, sod homes were constructed by some of the families who sought to make a living on the prairie. About half an acre of ground was turned over and the sod was cut into bricks about three feet long. The house was set into the ground and the roof covered with sheets of brush or tar paper. Although the sod house kept the family warm in the wintertime, it was quite leaky when the snow began to melt or when it rained.

Today in some of our most densely populated urban areas such as Minneapolis, Minnesota and Edmonton, Sascatchewan (Canada) to name just two, there is a developing trend to burrow into the ground rather than tower over it. The modern builder of an underground shelter has mated the concept of the cave to modern technology.

For a structure that is surrounded by the earth on all sides, the earth becomes the insulation. By directing the primary opening toward the south, the modern cave dweller has few requirements for heat. Unlike conventional houses exposed to ranges of temperature fluctuation from freezing temperatures to those hot enough "to cook an egg on the sidewalk," the earth-sheltered house is surrounded by a constant earth temperature of around 45 degrees Fahrenheit.

To bring the house to a comfortable temperature, say in the 60s, would therefore only require a modest addition of heat. Warmth from the sun can be obtained directly from south-facing windows or from a clear dome located directly over the rooms themselves. The rest of the necessary heat can be provided by a modestly sized heat source. (If, for example, the ground temperature is 45 degrees Fahrenheit and the outside air is 0 degrees Fahrenheit, the earth-sheltered house would require only a 20-degree rise in temperature to maintain a comfortable interior temperature of 65 degrees Fahrenheit. To maintain the same temperature in a conventional house with 0-degree air surrounding it, a 65-degree rise in temperature would be necessary.

David Wright, in *Natural Solar Architecture,* sketches a picture of an earth dwelling of the future: "A life-support pod suitable for any climate." The house would include air tubes, photovoltaic cells and an atmospheric static precipitator, with all heat and lighting functions controlled by computer. All wastes would be purified or converted, and all water would be recycled. Growth of all vegetables and protein foods would occur in growth pods and hydroponic tanks [1]. Certainly by Wright's accounting, a movement *back* to the cave seems more like forward progress.

Author's note: If the idea of an earth shelter seems appealing, advocates of the underground movement remind us that despite its historic connections, the earthbound structure is still rather "unconventional" in banking terms. Thus if one views it simply as a concrete or block structure with additional waterproofing and insulation and dirt piled up around it, it may find broader acceptance if a loan is needed.

GREEK DWELLINGS

Adaptations of historic dwellings and heating methods to modern society abound. Much of our own civilization is based on ancient Greek civilization. If we were to turn to Greek architecture for ideas for living spaces, we would quickly call to mind the classic images of their most famous structures: the Acropolis, and the Parthenon. The agora, the building that was centermost in many of the small Greek city-states served a community function much like the village shade tree.

Yet, upon closer study of Greek houses we would see that, in essence, they were nondescript. The exterior walls were strictly utilitarian. Whatever windows were present were usually placed high on the wall, in part to provide a measure of privacy in a densely-settled area. From the narrow streets (some were no wider than stairways and many, in fact, were stairways) the entries to the houses were like openings leading into a mouse-hole. The uninviting uniformity, however, hid the fact that an interior wall in many Greek homes opened to a protected inner courtyard.

From the drawings of city-scapes of the time, a definite intention to align the streets to form a predetermined grid seems apparent. When the city of Periaeus was rebuilt in the 4th Century A.D., for example, a framework of intersecting streets running north to south and east to west was set down so that all homes lining the streets might have at least one wall facing south for access to the winter sun [2].

By studying the city maps further, one can determine that shade trees—perhaps because of the narrowness of the streets—were usually confined to the areas surrounding statues or monuments or to the outer boundaries of the city. Because there were no trees shading the houses, access to the sun's warmth was ensured. And the intense clustering of houses and the surrounding wall barred any serious frontal attack by the wind.

It is interesting, in reviewing how the Greeks structured their cities, to remind ourselves that the concept of mandatory solar access (that is, requiring that new streets and new dwellings take into account a proper orientation to the sun) is far from accepted in the American culture. For a country that prides itself on its technical accomplishments, we have, time and again, attempted to thwart the natural course of events. We have let technology make the decisions for us, building to challenge nature instead of complement it. However, such a simple mechanism as requiring that all new streets be aligned to an east-west, north-south grid to enable direct solar gain by each new house would ensure that as much as 30 percent of the heat supplied to each home could come from a natural, freely supplied fuel source: the sun[3].

A ROMAN DESIGN

When in ancient Rome, one did as the Romans did, and went to the daily baths. Although aspects of the bath were described earlier in the book, one important feature was not mentioned which has been adapted for today's passive solar structures. The subterranean wood-fed furnaces that heated the water for the hot baths also dispersed hot air throughout the buildings. The heat traveled in hollow corridors beneath the floors of the bathing rooms and up the air shafts between the inner and outer building walls. Though at first glance this method for dispersing heated air may seem similar to our modern central heating system, it was not. The Roman hot air system did not mix heated air with the cooler air in the bathing rooms. Rather, the heat was absorbed by the floor and walls and thus released into the rooms.

Frank Lloyd Wright and others experimented with passing electric wires and utility pipes beneath floors and walls and even above the ceiling. However, because masonry heats slowly, such a method of heating assumes that the heat is not needed quickly. In addition, the demand for heating must be continuous, for once the heat buildup has begun, heat will continue to be released unless other measures—such as opening a vent—dissipate it.

The same heating principle used in the Roman baths has come into popularity with the development of the passive solar "envelope house." However, in this modern adaptation, the heat rises not from a furnace but from solar collecting panels or an attached greenhouse. The rising, warmed air is drawn by convection into the house and passes beneath the attic floor toward the north wall of the house. The air, cooling somewhat, is then drawn down the double-walled north side and lingers in the small air space provided beneath the house. The air space can also include rock storage for long-term heat retention. Because this space is open on the south end to the greenhouse, the convective loop is completed—as the air in the greenhouse is warmed, it will draw the cooler air out from under the house. The cycle can be broken by simply opening vents in the greenhouse or the attic [4].

THE CHINESE CONNECTION

Since the dawn of Chinese civilization, the practical Chinese have not pursued diversity in their building structures. Rather they have assimilated and refined their building techniques over the past 2500 years in a structural design most commonly known to us as the pagoda style. With such single-minded pursuit, they have achieved a sort of continuing perfection of a single, preferred style. Except for variations due to terrain, the same building design is applied to all domestic, official and religious architecture.

Over the centuries, the Chinese have drawn upon natural principles of heating and cooling. Their building style, which ultimately spread to Korea, Japan and elsewhere, consists of a timber framework placed upon a masonry (heat storage) platform. Because the framework construction places the weight of the building around the outer edges instead of in the walls themselves (as in conventional European and American dwellings), the walls can serve as spaces to admit or exclude light and air as desired. Because they do not support the building structure, they permit a flexibility not found in typical contemporary styles.

To gain as much heat as possible from the sun, buildings in China are usually oriented toward the south. As in our modern solar homes, the maximum amount of winter sunlight can then be drawn into the house.

The familiar, gracefully curving roofline that extends beyond the outer walls of the Chinese house, like the curling petals of a rose, serves more than a decorative function. The overhanging eaves, while allowing the low-angled winter sun to shine on the building, exclude the sun in the summertime when it is high overhead.

The Japanese, who have adapted the pagoda style to their own culture, use a more moderately sloping roof. However, not only does the Japanese building have a special overhanging eave (usually from one and one-half to three feet), but every doorway and every window has a special projecting roof as well.

When more light is desired in the building, windows or doorways can easily be cut into the walls of the timber frame house without damaging the integrity of the building structure. Because the interior walls are also not part of the supportive structure, they can be added or subtracted as demands for interior space change [5,6].

NEW ENGLAND TRADITIONAL

As mentioned earlier, the art of timber framework had been the focus of Chinese architecture for several thousand years.

Continuously evolving and refined over the years, timber frame construction became the dominant design for most buildings in China as well as in other nations, such as Japan. However, timber frame construction was also prevalent on the European continent and in colonial America.

For many people the term *timber framing* immediately, and erroneously, brings to mind the image of a log cabin. In Finland, for example, the first formal structures for shelter in a country abounding in timber were made of wood. Tree trunks, hand-hewn from the nearby forests were stacked one on top of another like building blocks. But these sturdy structures did not permit the flexibility inherent in the timber frame.

In Europe, timber-framed houses were built in great numbers between the 14th and 16th Centuries. The most familiar of these were the picturesque herringbone, square-panelled or close-studding-styled timber frames with wattle and daub (woven stick and mud) plaster. However, in the 16th Century, the use of wood for home construction came to an abrupt halt and stone and masonry were used instead. At that time there was also a dramatic change in the way the English heated their homes. Both changes were brought about the by same development. With its wooden naval fleets and burgeoning, wood-fueled industries, England was facing a serious wood shortage. The price of firewood, for example, increased over 800 percent between 1531 and 1632. This dramatic price rise forced the British to turn from their love of the open fire and begin heating with lumps of "strange smelling" coal.

Because coal was more plentiful and could be had for a much more reasonable price, the English adjusted by adapting their fireplaces, making them smaller and more shallow so that they could accept grates for burning coal. Although the woodstove was coming into widespread use on the continent, offering far more heat with less fuel, the English persisted in using the inefficient fireplace, despite the scanty amount of heat it provided.

For the colonial American family seeking a solid shelter from winter storms, a timber frame house such as the traditional New England Saltbox could be constructed from wood felled in nearby forests.

However, while New England-style timber frame houses were built to stand for centuries, this method of construction fell into obscurity by the mid-20th century. A major reason was that the homebuilding market, responding to the availability of cheap fuels for home heating, turned to using lighter weight, prefab materials that could be milled and transported great distances to the construction site.

The relative anonymity among neighbors in urban settlements was a problem too. In the past a person building his timber frame house could depend on the good will and labor of a community of neighbors to help raise the frame of the house. Often a houseraising could be accomplished in a single day.

By the early 1970s, although the timber frame tradition had been carried to America by the early colonists, the use of words associated with heavy timber frame houses had all but disappeared from the working vocabulary of most Americans. Verging on the brink of extinction in house building were words like dovetail, tenon and mortise. Such terms referred to methods of construction used in timber framing, methods that required painstaking attention to detail to create a durable structure that would provide shelter for hundreds of years.

Today, this classic style of building is undergoing a revival because of the escalating costs of both energy and housing materials.

The New England Saltbox [7], in particular, provides an excellent example of the assimilation of old and new heating techniques. By tradition, the roof of the Saltbox is set at a steep pitch to provide a snow shedding function that is essential in parts of the country where snow buildup is a problem. But the roofline also provides a suitable angle for the installing of a modern array of solar collectors. Assuming proper orientation of the house toward the south, the collectors can thus provide additional warmth for space heating and for hot water.

The Saltbox, with its windowless northern wall, designed to serve as a buffer against cold northerly winds, was doubly protected in colonial times by the addition of a storage shed that ran the full length of the northern wall. Today the northern side can be constructed with both an inner and outer wall. With solar collectors or a greenhouse on the south wall, the Saltbox style is as heat-efficient as the "envelope" style briefly described in the section on Roman designs.

As with the Chinese timber frame, the floor design of New England-style construction provides for easy partitioning of space. Because the weight of the structure is on the outer supporting timbers and not on the walls themselves, the exterior walls can easily be redesigned to let in more sunlight [8-10].

* * *

As these brief historic illustrations have attempted to suggest, the past, present and future are with us at any moment we become aware of them—with one exception:

Through use of the central heating system man has attempted to substitute technology for natural systems. . .breaking with a tradition as old as civilization itself. Only by assuming an infinite supply of fossil fuels to enable us to pump invisible heat into (and out of) every corner of our houses have we allowed ourselves to be placed in a vulnerable position—one that would have us acting in desperation should our system fail us or should the fuel be no longer as affordable or as available as we would like.

The designs for change are with us now. But we must allow ourselves to adapt to change, to welcome the demands it places on our creative imagination.

I shall conclude with a story that describes the kind of adaptability to change that I mean. The story, which perhaps by now has become a legend, is told about an old Vermont farmer who was hard at work mending his fence, which ran alongside a county road, when he was approached by a road surveyor. The surveyor informed the farmer that after deliberate measurement he had determined that the farmer's land actually lay on the New Hampshire side of the state line. The old farmer leaned up against the fence and thought for a moment. Then his eyes filled with good humor as he turned to the young man and exclaimed, "Well thank the God Almighty! I couldn't have stood another of them Vermont winters!"

REFERENCES AND NOTES

1. David Wright, *Natural Solar Architecture* (New York: Van Nostrand Reinhold Co., 1978).
2. R.E. Wycherley, *How the Greeks Built Cities* (London: Macmillan, 1967).
3. Ken Butti, "2500 Years of Solar Architecture and Technology," *Co-Evolution Quarterly,* Winter 79-80. (Sausalito, CA)
4. National Geographic Society, *Everyday Life in Ancient Times,* Washington D.C. (ca. 1947).
5. "China: Arts, Language, and Mass Media", *Encyclopedia Americana,* Vol. 6, 1979.
6. Marcus J. Sims, *The Japanese House and Garden* (New York: Praeger Publishers, 1955).
7. Stewart Elliott and Allen Harlow, *The Primer*: A Comprehensive Introduction to the Solar Starr House. . .and Blueprints, Boulder, CO: Dovetail Press, 1979). (P.O. Box 1496, Boulder, CO 80306).

8. Hans J. Becker and Wolfram S. Scholote, *New Housing in Final* (London: Alec Tirante, 1964).
9. David W. Brown, *Keeping Warm in New England,* (thesis, Master of Architecture) Massachusetts Institute of Technology, 1976.
10. Lawrence Wright, *Home Fires Burning* (New York: Hillary House Publishers, Ltd., 1964).

About the Author

Alexis Parks lives in a sun-filled stone and wood house with her two children and kindly dog, Chianti, in the mountains west of Boulder, Colorado. She has written widely in the environmental and conservation fields for both newspapers and magazines, including *The Washington Post, Christian Science Monitor, The Nation, The Progressive, Mariah-Outside,* and *Wilderness Camping.*

From her experiences as director of the Boulder Office of Energy Conservation and out of the Governor's Office of Energy Conservation, where she served as assistant director of the State of Colorado's Energy Extension programs, she has gathered a broad knowledge of conservation needs and techniques. At present she writes a newspaper column "Energy Hotline," is in the process of writing another book, and is serving as director of Urban Energy Associates, Boulder.